Product Development

Product Development

An Engineer's Guide to Business Considerations,
Real-World Product Testing, and Launch

David V. Tennant
Windward Consulting Group, Virginia, USA

WILEY

Registered Office
John Wiley & Sons, Inc., 111 River Street, Hoboken, NJ 07030, USA

Editorial Office
111 River Street, Hoboken, NJ 07030, USA

For details of our global editorial offices, customer services, and more information about Wiley products visit us at www.wiley.com.

Wiley also publishes its books in a variety of electronic formats and by print-on-demand. Some content that appears in standard print versions of this book may not be available in other formats.

A catalogue record for this book is available from the Library of Congress

Hardback ISBN: 9781119780137; ePDF ISBN: 9781119780175; ePub ISBN: 9781119780182; oBook ISBN: 9781119780205

Cover image: © TUMBARTSEV/Shutterstock
Cover design: Wiley

Set in 9.5/12.5pt STIXTwo Text by Integra Software Services Pvt. Ltd, Pondicherry, India

SKY10033726_031122

Table of Contents

Acknowledgments

No one person can succeed on their own. There are always people that help along the way. Consequently, I have had the pleasure of meeting and speaking with a number of exceptional and knowledgeable people during the creation of this book.

First of all, kudos and thanks to Gary Elliott, President, International Applied Engineering. Over the years, Gary has provided significant insight into alternative energy systems, along with engineering and business perspectives in what was, a few years ago, a niche energy market. Gary continues to provide his expertise in clean energy power generation.

Thanks also to Tim Jarrell, Vice President of Power Supply, Rates and Distributed Generation at Cobb Electric Membership Corporation, for his thoughts on electric service and the future of electric trucks and cars relative to costs and meeting electric demand.

Along those same lines, David Jaskolski, Advanced Technology Vehicle Sales, Peach State Truck Centers, provided a wealth of information about large electric trucks, their applications, use, and future technology. David is truly a technical expert in this arena as well as understanding the commercial and economic issues in electric truck operations.

David Malone, Vice President of Marketing at Gas South, provided strategies for obtaining customer preferences and the use of digital marketing. Further strategy in this energy market was also provided by Kevin Greiner, CEO of Gas South.

Steven Lustig, Vice President of Global Supply Chain at East–West Manufacturing, was instrumental in discussing manufacturing techniques, remote teams, logistics, and global purchasing.

Roy Sequeira, Computer Systems Consultant, provided his perspective in developing and launching computer systems and products. His knowledge on planning and avoiding problems was most helpful; especially his observations on why launches sometimes fail.

Gerard Hill (retired), former Vice President of Consulting at ESI International and later an independent consultant. Gerard, over the years, has mentored and assisted me in project management techniques, methodologies, and practical applications.

Finally, to my wife and family who persevered through my absence while working on this book. Thank you for your patience and understanding.

I want to express my sincere thanks and admiration to each of the above participants. All of today's knowledge is built on the experience and knowledge of previous pioneers. These gentlemen represent the leaders in today's new thinking and product development.

About the Author

David V. Tennant has directed over $3.5 billion in projects, programs, and resources. His expertise is in energy, utilities, telecom, manufacturing, and consulting. He has had engagements worldwide and currently resides in the Atlanta area.

His 30 plus years of functional experience spans engineering, operations, marketing, executive management, consulting, and he currently serves on two corporate boards.

He has a B.S. in Mechanical Engineering from Florida Atlantic University, a M.S. in Technology and Science Policy from the Georgia Institute of Technology, and an EMBA from Kennesaw State University. He is a registered professional engineer (PE) and a certified project management professional (PMP).

His LinkedIn profile can be found at: www.linkedin.com/in/david-tennant-57075.

1

Introduction to Product Development

My formal academic training began as a student of mechanical engineering. Simultaneously, I worked part-time as a designer to pay for my education. In this case, my work allowed me to see how products were developed to the lower level where I was doing the technical drawing. And my engineering education helped me learn the limits of materials, think about product technical features, and how to apply mathematical formulas to solve technical problems.

However, nowhere did I learn about how products are *really* developed. What drives a company to success? How do companies know which products will be accepted in the marketplace? What is a marketplace or a market segment? As a result, my education and early work experience taught me a lot about applying engineering principles, but I had no knowledge of marketing, sales, business finance, C-level executive support, or how R&D (research and development) and the other areas are supposed to all work together. I simply (and naively) believed that new products were developed and launched by the engineering department. The product that had the better design would always be preferred by consumers.

Since those days in college, and my early career in engineering, I have come a long way to understanding that product development is a multi-faceted effort involving many diverse groups and talents. It is so much more than R&D or engineering.

Similarly, I have noticed that other disciplines – marketing, sales, etc. – do not always understand the engineering or R&D process in product development, which can be equally frustrating for those on the business side of a company.

The underlying purpose of this book is to serve as a bridge between the various groups that are responsible for developing and launching new products and services. It will assist engineering students, marketing professionals, R&D scientists, and product developers in how to effectively plan and launch new products.

This does not mean your future projects will all be successful: the market determines who survives and who doesn't. However, the techniques, methods, and studies in this book can help you determine which projects should go forward and how to best plan and execute a flawless development and launch. And which projects should be scrapped early in the development process. Half of the key is knowing which projects have a greater chance of success vs. those that don't. To illustrate important points, case studies will be provided that illustrate real-life failures and successes that

Product Development: An Engineer's Guide to Business Considerations, Real-World Product Testing, and Launch, First Edition. David V. Tennant.
© 2022 John Wiley & Sons, Inc. Published 2022 by John Wiley & Sons, Inc.

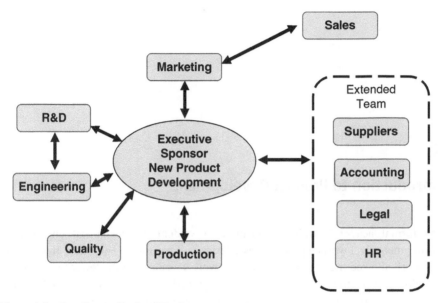

Figure 1.1   Core Team – Product Development.

reinforce the principles in this book. This book is very practical and does not dwell on theory or untested ideas.

Figure 1.1, Core Team in Product Development, illustrates some of the various functional areas and their interaction in product development. This figure is dynamic, so the participants will change depending on the type of product, its complexity, and it's corporate visibility or priority.

It should also be noted that product development will require market research, data analytics, realistic objectives, project management, and leadership.

Project Management and Product Development

A part of this book will introduce the reader to key concepts of project management which can be useful in bringing a product to market in a timely and efficient manner. Of course, the project manager, or leader, plays a key role in planning how the work will be accomplished, how it will be executed, and by whom. Further, this book will provide insights and processes in developing new products. It is not a book about project management; however, because project management (PM) is such a core skill needed to successfully launch new products, the reader will find key PM concepts detailed throughout the book. This will offer a good grasp on how to manage a project using these techniques and the reader is encouraged to pursue further reading on the topic of project management. It will be apparent that "product management" and "project management" are used interchangeably.

It should be noted that product development requires a multi-discipline approach, with each participating department understanding their role and responsibility. Many times, people have described PM as similar to herding cats. Product development will have similar challenges.

Through research, my own experience, and discussions with product development professionals, it is clear that project management plays a strong role in staying focused on the scope, budget, and schedule for the new product's development. Successful companies have recognized over the last 20 years that project management processes are a huge competitive advantage. It has always been the author's perspective that successful product launches require two ingredients: technical excellence and managerial excellence.

In recent years, there has been a drive to embrace an agile project management philosophy. This is much more suited to software and IT projects and certainly has its place in that industry. The primary differences between traditional and agile project management will be presented in Chapter 5.

What Is Product Development?

We first need to ask ourselves, what is a product? A product is simply an item or service that brings value to the customer in exchange for a price. You may find it odd that a "service" can be considered a product, but it is. A service can be a utility providing electricity to its customers. It can be an insurance policy that you purchase for your car, or the lawn service that keeps your company's green spaces looking nice. Services can also be developed and sold similar to "hard" products. While services can be considered a product, marketing services is different from marketing a hard product. Further, think about apps that are developed for your smartphone. Are these a product or a service? Sometimes the line can be blurry.

It is appropriate to consider that products can originate from a variety of sources. Perhaps your R&D group has developed a new smart widget. It is most likely now the engineering department's mission to determine how to commercialize it and manufacture it in a cost-effective manner.

The marketing group will need to determine the channels necessary to distribute the widget, how to develop a sales and marketing strategy, and determine estimates for pricing, profit, breakeven point, etc. Essentially, there are many moving parts to successfully launching a product. Where do products come from? Figure 1.2 shows some (but not all) of the points whereby ideas for new products are generated.

How This Book Is Organized

It is suggested the reader go through each chapter in sequence. This is because each chapter builds on concepts presented in the previous chapters. Also, developing products should have a structured approach and this book provides this in the correct order.

In many chapters, there are "Tabletop Discussion" questions. These are provided for group discussions and should have about a 30-minute time frame. These should encourage students or product teams thinking about key topics and how to address them.

Case studies are included to reinforce the concepts presented in each chapter. While there are one or two "hypothetical" cases, most of the cases presented are real-world

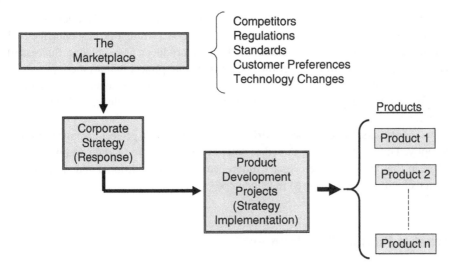

Figure 1.2 Where Do Products Come From?

examples. The author has had significant experience in rescuing projects (or product developments) that were in deep trouble. In these instances, it would be in poor taste to name companies or individuals. Many of these cases challenge the reader with questions pertaining to how they would have addressed, prevented, or rescued these difficult issues.

Key points are summarized at the end of each chapter, followed by questions for students or readers to answer.

The answers to both the case studies and questions are also provided.

Subsequent chapters include:

Chapter 2 – The Role of Marketing

This provides an overall perspective on the basics of marketing, how it is different from sales, and how pricing, product, promotion, and place are part of a strategy. Marketing many times drives new products; indeed, a marketing professional may be placed in charge of a product's development. Marketing also performs competitor analysis, obtains customer feedback (e.g., focus groups), and develops sophisticated models – usually an Excel spreadsheet – to evaluate pricing sensitivity.

Chapter 3 – The Role of Engineering

Engineers play a key role in the development of products. And there are many disciplines in this field. Listed in Chapter 3 are the types of engineering – mechanical, electrical, instrumentation, etc. – and how they work together to develop a project. There is a controlled approach to defining and solving engineering problems. Also, computer modeling can reduce development time and reduce the number of prototypes that are produced.

In addition to the technical challenges, engineers often have to design a product with the potential to be misused, either accidentally or intentionally. This brings into focus the need to consider ergonomics (human factors engineering) and product liability issues.

Chapter 4 – Core Team and Teamwork

Which groups and who should be involved in the product's development? What is the role of executive management in the process? These questions are evaluated from the perspective that a team is required. How do teams work together to be most effective? The use of matrixed teams – so common in many large companies – can be positive or detrimental depending on the organization. Teamwork and leadership relative to product development will be explored.

This chapter will also look at the role of the accounting, finance, and supply chain (procurement) departments. Having a product that continuously goes over budget will affect its profitability and have the potential for product cancellation, if left unchecked. This chapter is important as it clearly defines the roles of these functional areas and their contribution to a product's development.

Chapter 5 – Getting Started

Once a product/project has been approved by senior management, how do we get started? The reader will learn how the business case, typically a marketing function or feasibility study, is the springboard to developing an action plan. Also, the differences between basic and applied research will be presented. How does R&D (research and development) participate in a product's development?

The concepts of formal project management (and several case studies) will be offered along with the top reasons that projects, or product developments, fail. Both agile and traditional project management theories will be presented and understanding when to use them.

Chapter 6 – Product Development for Small Firms and Entrepreneurs

Small firms have a set of challenges that are unique. Many times, obtaining credit, business loans, or investors is a challenge. This is not something a larger, well-established firm is concerned with. However, this chapter delves into sources of funding (venture capital, angel investors, etc.) and where to find local incubators, which are set up to assist small companies and inventors with skills, training, and business contacts.

Many times, small firms or inventors do not have access to talent such as engineering or manufacturing companies. These are discussed along with product roadmaps and when or if to obtain a patent.

Chapter 7 – Manufacturing the New Product

A history of Japan's focus on quality and manufacturing is explored along with a history of America's car companies from the 1960s. Technology, government policies, foreign competition, and other factors can impact design and manufacturing. Just-in-time and lean manufacturing are presented with the benefits and drawbacks to each.

A review of small vs. large manufacturers and domestic vs. offshore will be offered. Further, an overview of current manufacturing techniques will be offered as well as a discussion on what manufacturing will look like in the future: 5 G, artificial intelligence, the Internet of Things, and robotics.

Chapter 8 – Engineering Product Design and Testing

The project lifecycle is introduced along with methods to perform risk reviews. A risk is simply a potential, future problem that can ruin your project. A risk review can assist in identifying problems and developing mitigation strategies.

A discussion and an example of engineering modeling is presented. This illustrates how modeling saves time and can be very reliable. Supply chain is equally important as it must work closely with engineering in issuing RFPs and technical specifications. The chapter closes with a review of new technologies and the importance of identifying stakeholders.

Chapter 9 – Successful Product Launch and Post Review

A review of failed products is presented along with commentary about their demise. This chapter also complements Chapter 2 (Marketing) as launching a product has significant implications for strategy, pricing, sales partners, and positioning.

A post review is a useful exercise to determine what went right in a product's design, production, and launch for future product development teams. Equally important in the review is what could have been done better? Both sides offer opportunities for continuous improvement and avoidance of repeating mistakes.

Chapter 10 – Summary

All of the previous chapters have dealt with specific topics and are somewhat siloed. However, this chapter will connect all of the dots so that a structured approach to product development can occur. A process flow diagram will demonstrate the activities that occur from beginning to end.

Takeaways from this book.

- Product development has substantial costs and risks, but also high potential.
- Both managerial and technical excellence are required for success.
- Successful product development requires the use of project management techniques.
- The world does not revolve around engineering, science, marketing, or IT, it revolves around profitability.
- Product development is a team effort generally requiring talent from different functional areas.
- Product development is often led by a senior-level marketing person. Other times, it may be someone from R&D or engineering.
- Bringing a product to launch within schedule and budget is very important. If late, competitors may be first to market or have a better product. If costs spiral out of control, it may take years to recover the investment costs and the product may never be profitable.
- Engineering modeling can significantly reduce product development time and minimize the need for prototypes.
- As a product goes through development, ergonomics (human engineering) should be a design factor.
- Products should be designed assuming they will be misused – either by accident or intentionally – so that possible product liability can be minimized.

2

The Role of Marketing in Product Development

Corporate Strategy – Strategic Planning

What is strategic planning? This can mean different things to different people. While strategy can be used to guide a company to the next level of performance or change the corporate culture, in product development, strategy can help bring focus. Even corporations get operational plans (short-term focus) mixed up with strategic plans (long-term focus). Essentially, the strategic plan is performed at the highest levels of corporate leadership and may even involve the firm's Board of Directors.

A strategic plan (corporate strategy) should focus on long-term actions to move the company in a new direction. For years, Coca-Cola was primarily distributing its iconic soft drinks through distributors all over the world. However, as people became more concerned about sugar and corn syrup content in soft drinks, Coke, and other drink makers, decided as a strategic decision to augment their portfolio of products. This now includes Dasani® bottled water, Minute Maid® fruit juices, as well as several tea and coffee products. A conscious decision (or strategy) was made to diversify Coke's product mix to better serve their customers' tastes. Their largest competitor, namely PepsiCo, has followed the same strategy and offers not only a variety of beverages, but also various food lines, including Frito-Lay® and Quaker Oats®.

For technology companies, there is a similar approach to offer a continuous stream of new products. If we look at Apple, their initial offering was their innovative computers and operating system. But the firm has evolved to offer a plethora of new and different products. At one point, Apple was bringing in more revenue with iTunes than with their computer business. Added to the mix are the iPhone and iPad. Clearly, their corporate strategy has included diversification and innovation, and is driving the firm to develop new products and services. It is important to note that corporate strategy will drive innovation and provide the funding for R&D, which also includes funding for new products.

Figure 2.1 Where Do Ideas Come From? suggests that companies have many avenues to generate ideas, both internal and external. Usually, anyone within a corporation can offer suggestions and ideas. At some point, the ideas will percolate to senior management, who will decide the merits of each suggestion and whether to direct company resources (people and dollars) in this direction.

Product Development: An Engineer's Guide to Business Considerations, Real-World Product Testing, and Launch, First Edition. David V. Tennant.
© 2022 John Wiley & Sons, Inc. Published 2022 by John Wiley & Sons, Inc.

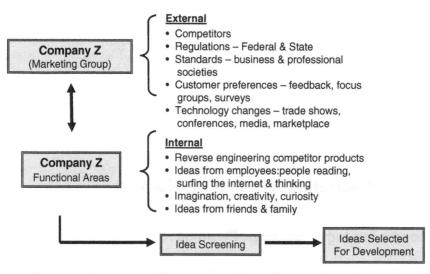

Figure 2.1 Where Do Ideas Come From? Figure by David Tennant

This is related to innovation. How do we define innovation? Here is a concise definition:

> Innovation is a process by which a domain, a product, or service is renewed and brought up to date by applying new processes, introducing new techniques, or establishing successful ideas to create new value.[1]

Innovation is what drives new products. How do we innovate? Innovation implies that your firm has a creative spark, or the founder has an idea that no one has yet thought of. Innovation is not something you can learn in school or from a co-worker. Innovation comes from creative people.

Many times, innovation comes from someone being aware of a market direction or trend. Others may think about how technology changes can be incorporated into a new product or software application. Also, new regulatory rulings can fuel a market need for compliance. True innovators are rare (think Nikola Tesla, Elon Musk, or Richard Branson – they all started from scratch) but working together with a partner or team of people can create synergies that, combined, can develop interesting new ideas or even an advancement of an existing idea.

All companies need to exhibit some form of innovation to survive. This is sometimes driven by the company founder, who generates ideas non-stop that have not been thought of before. We can also think of innovation as a form of creativity or imagination, but this is not something we can learn in school or by taking a course. Some people are just more creative than others.

It must be noted that innovation and new products are the future profits and livelihood of the corporation. A company cannot sit on its laurels, it must continue to generate new products. If one looks at a list of the top US companies (Forbes or similar) 40 years ago and compare this list with today's top companies, you will find the list is very dynamic and different. Table 2.1 shows the Fortune 500 list of companies for 1980 and 2020. They are listed by revenue; however, column three is listed by profitability, which is different.

Table 2.1 Largesr US Companies 1980 vs. 2020 Revenue.[1]

Rank	Company[2] 1980	1980 Revenue	Company[3] 2020	2020 Revenue	Most Profitable[4] 2020
1	Exxon-Mobil	$ 79.1	Wal-Mart	$ 524.0	Apple
2	Gen. Motors	$ 66.3	Amazon	$ 281.0	Microsoft
3	Mobil	$ 44.7	Exxon-Mobil	$ 265.0	Berkshire Hathaway
4	Ford	$ 43.5	Apple	$ 260.0	Alphabet (Google)
5	Texaco	$ 38.4	CVS Health	$ 257.0	Intel
6	Chevron-Texaco	$ 29.9	Berkshire Hathaway	$ 255.0	Facebook
7	Gulf Oil	$ 23.9	United Healthcare	$ 242.0	Johnson and Johnson
8	IBM	$ 22.8	McKesson	$ 214.0	Verizon
9	Gen. Electric	$ 22.4	AT&T	$ 181.0	Pfizer
10	Amoco	$ 18.6	Amerisource Bergen	$ 180.0	Wal-Mart

Table Developed by David Tennant.
Numbers in billions of dollars

Note that some of the companies that were on the list in 1980 are now absent. This is especially notable for the oil companies, and this will continue as the shift from petroleum (gasoline) to electric and alternative-fueled vehicles continues.

New firms in the form of big tech, Facebook, Google, etc. have been displacing more "traditional" companies. By the 2060s we may find a whole new set of innovative companies on the list.

At the turn of the last century (1900) the horse and carriage gave way to the automobile. There were most likely excellent buggy whips for sale, but no matter how excellent your product may be, innovation will ultimately replace it.

If we consider Kodak, which had the lion's share of the film market, this was completely made irrelevant by digital photography. The major camera producers: Canon, Nikon, Olympus, and others developed highly successful digital cameras which took excellent pictures and no longer required film. Apple's iPhone further hastened the decline of film in that anyone could take very good pictures with their phone's camera (including Android phones).

Discussion Case 2.1 – The Eastman Kodak Company Based on[5]

The Eastman Kodak Company, founded by George Eastman, devised and marketed the first simple camera in 1888. As time went on, Eastman incorporated the name Kodak as part of his product line, which included various film products (Kodachrome, Ektachrome, etc.). Later, his company turned out 35 mm format slides, 8 mm and 16 mm movie film, and a variety of projectors and cameras, for both professional and personal use.

Continuing with innovative and interesting products, Kodak also produced slide projectors, the Browning personal movie camera, and film-processing machines. Many Hollywood movie studios used Kodak film in their movies. By 1976, Kodak had 90% of the US camera film market. The peak revenue year for Kodak was 1996, with revenue of $16 billion. The company was traded on the stock exchange with a share price of $90. At that time, Kodak had two-thirds of the global film market. Employees peaked in 1988 with over 145,000 employees worldwide.

Kodak continued in their attempts to be innovative with digital cameras and the CD format for pictures, but each camera sold lost $60. In 2004, it was ejected from the Dow Jones stock exchange. As a side note, the Dow Jones Industrial Average is a stock index of the top 30 US corporations. On this note, Kodak closed many manufacturing plants and cut thousands of jobs.

Kodak worked to restructure its business, acquired several companies, and provided licensing which brought in much-needed revenue. In 2010, Kodak sued Apple over its iPhone technology, claiming a patent infringement, but did not prevail. By this point, employment was down to 18,000 employees. In 2012, the company filed for bankruptcy and sold its online photo business to Shutterfly.

Kodak was somewhat reluctant to move into digital photography as it did not involve the use of film. The company today provides traditional and digital printing, hardware, and software services to the print, packaging, publishing, manufacturing, entertainment, and film industries. The company's four primary segments include advanced materials and chemicals.

At the time of writing (August 2021) Kodak's stock price is $6.86 and the company has earnings of negative $400 million on revenue of $1.1 billion.[6]

Case 2.1 Discussion Questions

1. As a large, successful US company, Kodak was once a "dogs of the Dow" firm listed on this stock exchange. It could be argued that Kodak's management was unwilling to change or consider new ideas to displace or augment its line of film products. What does this tell us about the role of innovation in a company?
2. With the advantage of twenty-twenty hindsight, it is easy to criticize Kodak's lack of foresight or imagination. However, it is likely that past success blinded Kodak to the future of film. This might be categorized as a resistance to change, which is found in most large corporations; it becomes part of the corporate culture.

    If you are in a senior level position – at any company – how would you work to change the corporate culture?
3. What role should the marketing department have played in this long-term decline?
4. From a managerial standpoint, what are the challenges when a company is in decline?

Marketing, Sales, and the Four Ps

A common misunderstanding is "marketing is the same as sales." They both are concerned with increasing the company's sales. Both need to be knowledgeable about who their competitors are and their products. Both need to understand their current and potential customers' needs. But there is a difference. Let's first consider the sales team.

The sales group is first and foremost responsible for bringing in revenue. A second objective is to establish relationships with customers. It is always easier to sell to an existing customer than to find a new one. However, to increase sales, a sales representative must also be able to meet potential new customers. This can be done in a variety of ways. The traditional methods include attending trade shows, cold-calling (i.e., working the phone), attending local business meetings (city business chapters, chambers of commerce, etc.), and attending regional or local professional trade associations; for example, engineering chapters such as ASME (American Society of Mechanical Engineers), IEEE (Institute of Electrical and Electronics Engineers), and other similar organizations. Beside technical groups, there are organized chapters of venture capitalists, human resources management, etc. Any professional group will be a source of potential customers.

In today's environment, where pandemics can disrupt traditional sales prospecting, most companies have a strong online presence. This is in addition to your company's website, and may include sites such as LinkedIn, Twitter, Reddit, Facebook, and a host of other platforms. However, an online sales presence starts to encroach into the domain of Marketing.

As noted earlier, the fundamental aim of sales is to sell products and obtain contracts: ultimately, to turn potential customers into current customers. Note that sales reps generally have goals to meet on a quarterly or yearly basis. This is typically a financial goal; for example, "Jill must bring in $400,000 dollars in revenue each quarter." While the sales rep is concerned about meeting his or her sales goal, it is also important that products and services should have enough profit (also called "margin") so the company makes enough money to stay in business – and ideally to grow.

The marketing group, while concerned with sales and profits, plays a very different, but complementary role to sales. Marketing is integral to the financial health of a company and must support the sales team. Pick up any book on marketing, and you will learn about the four Ps:

- Product
- Promotion
- Pricing
- Placement.

These are the marketing fundamentals, but true marketing can be very analytical, with spreadsheets, detailed market analyses, and estimated demand for the product. But first, let's focus on the 4 Ps.

The 1ˢᵗ P – Product

It is not unusual for a Marketing Manager (sometimes called a Product Manager) to have the leadership role in the development of products and services. This will become apparent as our discussion unfolds.

Obviously, to have sales a company needs to have a product that has demand in the marketplace. The market is constantly churning with established products being bumped by newer products. Perhaps the newer product is cheaper, has better features,

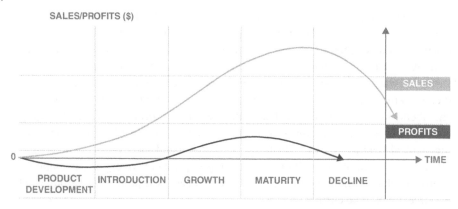

SALES/PROFITS ($)

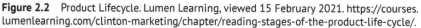

Figure 2.2 Product Lifecycle. Lumen Learning, viewed 15 February 2021. https://courses. lumenlearning.com/clinton-marketing/chapter/reading-stages-of-the-product-life-cycle/.

or has captured the imagination of the buying public. History is littered with products that were successful but ultimately replaced by a competitor.

All products exhibit a lifecycle. Figure 2.2 shows a typical product lifecycle. Notes about the product lifecycle:

Development stage –

- Product/project is funded, and design of prototype starts
- Prototype put through testing, and data collection
- Review of data; evaluation of results
- Changes in design and more testing
- Final product handed over to manufacturing or external group for production

Introduction stage –

- The product is new to the market and demand will increase with time
- The cost per unit (or customer) is high
- Few if any competitors
- Sales are low and the product may be a money loser during the introductory stage

Growth stage –

- Sales and revenue increase; the product is taking off in the marketplace
- Competitors are becoming aware of your activity and may start to develop a competing product
- Cost per unit (or customer) begins to decline
- Profit margins increase

Maturity stage –

- Sales, revenues, and profits peak
- Cost per unit (or customer) is at the lowest point
- Competitors are now actively nipping at your heels

Decline stage –

- Sales and revenue decline, leading to a decline in profit margin

- Cost per unit (or customer) is still low, but may rise with fewer units sold
- Competitors start to fall away. A few will remain, but with a reduced market there are less sales and therefore profits to share. The product may become stable with steady sales but will not see the heavy growth or revenues seen in the growth stage.

Example of Product Displacement

In the late 1990s, Nokia produced cell phones that were widely used by the public – worldwide. Nokia, headquartered in Helsinki, Finland appeared to have a dominant position in the market. The Nokia 3210 (Figure 2.3) was compact, had many features, was very rugged and ergonomically comfortable to hold and operate. As a result, around 150-million units were sold. But, based on the previous discussion, we know that dominance in the marketplace can be upset by newer technology or lower-priced models with the same features. Along came the Apple 2G (Figure 2.4), which gained

Figure 2.3   Nokia Phone. Isaac Smith / Unsplash

Figure 2.4   Apple iPhone. Michal Weidemann / Unsplash

immediate acceptance by the public. The iPhone 2G had numerous apps, used a color touchscreen, and could connect with the internet in addition to all the things that the Nokia did (texting, photos, etc.).

Some general characteristics of products include:

Features – nice displays (computer screens), user friendly operation (electric vehicles), the wow factor (new and innovative, nobody does this – Wow!), and a variety of models to fit various budgets (low, middle, and high-end). The Apple iPhone definitely had the "wow" factor. Thirty years later, there are now competitors to the Apple iPhone.

Specifications – a technical description or requirement of what the product can do. For example, if our new product is an electric motor for electric vehicles: the materials used to make the product (steel, stainless steel, etc.); the number of motor brushes and material; outputs –horsepower, torque; the range of temperatures for optimum operation, etc. If you pick up any standard-powered car brochure, specifications usually include items such as horsepower, torque, engine displacement, fuel requirements, etc. Specifications may sometimes be spelled out by the customer for special orders or custom applications.

Warranties and guarantees – will your company stand behind its new product?

Service Centers – what if I need service for my new scanner/printer? Where do I go? Is there a center nearby? Note: a service center can generate additional after-sale profits: think of automobile dealers and their service center networks.

What Marketing Cannot Do

The marketing group cannot guarantee that your product will be successful. Nor can marketing predict how many customers will buy your product, nor say with confidence how much revenue or profits will be gained. But spreadsheet models will be created that have various projections and sensitivity analysis. Marketing in today's world is very dependent on data analytics which helps clarify the target market and product acceptance. Table 2.2 shows an example of price, margin, and volume sensitivity analysis.

Table 2.2 Price and Margin Sensitivity for Solar Power Device.

10% of 16.9 million	Sale Price Price Sensitivity	TTL Revenue	Margin	Margin %
1,960,000	$ 50.00	$ 98,000,000.00	$ 24,500,000.00	25%
1,960,000	$ 45.00	$ 88,200,000.00	$ 19,404,000.00	22%
1,960,000	$ 40.00	$ 78,400,000.00	$ 15,680,000.00	20%
1,960,000	$ 35.00	$ 68,600,000.00	$ 11,662,000.00	17%

Table by David Tennant

Discussion Case 2.2 (Refer to Table 2.2)

Your marketing team has performed the following analysis. The new product is a portable external battery device that can power your laptop computer for up to one full day. The unit can be charged using regular home electric current or using the sun (solar power) in about one hour. This device is smaller than competing devices (a little larger than a flash drive) and more versatile. The target market is college students, nationwide. The new product has been designed to be trendy, easy to use, and reasonably priced. There are approximately 19.6 million college students in the USA.

We will assume that 10% of the total number of students will purchase the product and that the average cost will be $50. Note that the margin (profit) declines per unit as price declines. Discussion questions for Case 2.1:

1. What are the advantages and disadvantages of this kind of modeling?
2. Why do you think margins per unit decline as the price drops?
3. Looking at the sale prices compared to margins (%), is $50 the best price? How do you know students will pay this amount for your product?
4. If 10% of students purchase your product the best-case scenario shows $98 million in revenue and profits of $24.5 million. The worst case is a profit margin of $11.662 million. In your opinion, are these numbers realistic?
5. Do you think you will have competitors prior to reaching the maturity stage?
6. Is this type of analysis realistic?
7. What role would the engineering or R&D group play in this evaluation?
8. The chart above shows gross revenues and net profits. There is no mention of the cost of the product. What are some of the costs that would be involved? Can you determine from the chart above the total "costs" of the product?

The 2nd P – Promotion

The promotion of your new product is critical. And this is where marketing generally gets confused with sales. Promotion deals with how to get your product and its message out to the public, ideally in front of your competitors. Promotion is where marketing creates value to customers, or even the perception of value.

In promoting your new product, you must think about segmentation. For example, in the previous discussion topic, it was determined that the initial target market was college students. This is an example of segmentation, which is one of marketing's primary concerns. To whom will we direct our promotional efforts? Do you think the above device is attractive to more than college students (high school students, working professionals)?

Segmentation can assist tremendously with the promotion of your product. Segmentation allows one to identify groups of potential users of your product. Here are several ways we can segment the market (and there are probably more):

Geographic: different areas of the country may have different tastes. Heavy winter clothing in February is fine in the Northeastern United States, but buyers in Florida will not have much use for it, even in February. If a firm does business overseas, then preferences will be different in color, texture, and marketing. For example, certain colors in North America symbolize happiness or purity; those same colors in other countries have the opposite meaning. Doing business overseas means one must be familiar with that country's culture, history, ethnicity, and general preferences.

Demographic – Your target market may be grouped by gender, age, income, race, and other factors. For example, Ferrari considers high (very high)-income people their primary target. Tickets to professional sports venues can appeal to a variety of income groups. Box seats, or those closest to the home plate, will charge a premium. Lower cost seats will be in the upper reaches of a stadium, further away from the home plate. Everyone is going to the ball game, but where you wish to sit determines your cost.

Behavioral – Some groups of people will consider a product in a variety of perspectives. Some of these approaches will be similar. For example, everyone uses milk, but people have different perspectives. Those who are diet conscious may prefer to buy 2% (less fat); children might prefer chocolate milk, those who are vegan will prefer soy milk. Your target market can be divided into multiple subsets.

The 3rd P – Pricing

In most companies, it is the marketing group that determines pricing and margins using spreadsheets, pro forma models, market analyses, and other techniques. In many instances, new product development costs may involve significant capital spending. Capital spending usually involves depreciating the costs over time and detailed models projecting costs and revenues over time are common.

Several techniques that can assist in determining price can also include competitor analysis and focus groups. If you know what your competitor's charge for a similar product, you have a starting point. Can you charge less and still be profitable? Will your product be so innovative as to pass your competitor allowing you to charge more?

Customer focus groups can provide insight as to what people are willing to pay for your product. Generally, a focus group will consist of a dozen potential customers, selected by your company, that will be introduced to advance notice of your new product. A few typical questions for your firm to ask includes:

- What features would you want in this product?
- Does this product fit your budget at $ _____.?
- What are your initial impressions?
- Do you like this more than the product you use now?
- If adding more features, what would you be willing to pay?

While a focus group can uncover many customers' likes and dislikes, pricing is clearly one of those attributes. Further, a series of focus groups, usually four or five, will help present a clear picture of market acceptance, pricing, general appeal, etc. Similar to market segmentation, focus groups can also be segmented with different groups according to income, age, etc.

In product development, there will always be costs to recover. Table 2.3 provides a listing of costs expended during development. Note that this is a general guide and is not an all-inclusive list.

All of the above costs must be recovered and reflected in the price of your product over time. It is important that a company recovers its development costs plus a margin;

Table 2.3 Typical Costs Expended in Product Development.

• Quality – QAQC	• Labor
• Testing services or lab time	• Capital equipment
• Travel expenses	• Testing equipment
• Regulatory compliance	• Engineering
• Legal	• R&D
• Permitting	• Equipment rentals
• Raw materials	• Expendables
• Marketing and advertising	• Training
• Corporate overhead	

Table by David Tennant

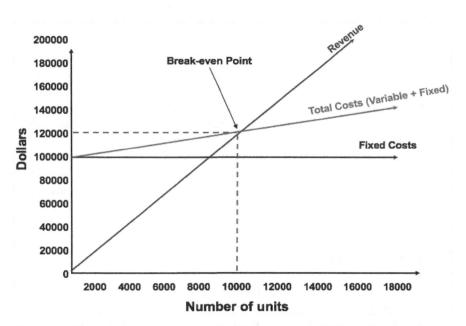

Figure 2.5 Breakeven Point. *Source:* Corporate Finance Institute (2021) / with permission from CFI Education Inc.

otherwise, your product is not profitable. It is likely that cost and pricing sensitivity analysis will determine the maximum amount of dollars that can be expended and remain profitable in a competitive marketplace.

Figure 2.5 shows a typical break-even graph for a product. Break-even is the point at which sales revenue equals all the costs expended; and additional sales will generate a profit going forward.

The 4ᵗʰ P – Placement

Placement is concerned with how and where your product will be positioned in the marketplace. As an example, Microsoft Office® is popular software used by individuals and corporations. Twenty years ago, one could purchase this product at any office supply store (Office Max, Staples, online, etc.) and many of the major retailers such as BestBuy, Sam's, and Costco. The product was a CD which downloaded the software onto your computer. Today, this product can be downloaded over the internet directly from Microsoft, BestBuy, Amazon, and other sites. Indeed, CDs have become technically obsolete as even music is now streamed or downloaded.

Another example is product placement in grocery stores. This is as much a science as an art. Grocery stores have significant research on customer buying habits and place products accordingly. For example, impulse buys such as candy, cold drinks, and general interest magazines are located in the checkout lines near the cashier. Wandering along the soft drink aisle, the major drinks are placed primarily at eye level to catch shoppers' attention. Some companies will pay for key product placement areas in the store. Finally, end-of-aisle displays are another location that customers always notice. Placement is important.

For software or services, how will this "product" be placed on the internet? This is where search engines can be useful. Google, Facebook and others have extensive market research from tracking internet users. This is how they make their money: selling market research, commonly known as data analytics. They can provide data on how many people visit key websites, perform searches for products, and can segment this data for companies willing to pay for it. When placing your product or service on the web, they can also help your firm target your demographic or geographic preferences.

Placement is key to getting your potential customers to notice your new product, whether a hard product of software based. The internet can help your new product get "placed" and increase your chances of success, assuming your product has appeal (demand).

The Business Case

The purpose of the business case (sometimes called a business plan) is to assist the executive team and Board of Directors make an informed decision. A business case is a detailed analysis of a new plan of action: proposing a merger or acquisition, developing a new product, or adding a new business unit to the company. It is generally a collaborative effort between several functional areas of the company, i.e., accounting, marketing, engineering, etc. In the technical or engineering world, it may be called a feasibility study, but the concepts are the same.

What is contained in a business case and why do we need one? Before a company invests millions of dollars into a product, merger, etc., it needs assurance that the venture will be profitable, makes sense from a strategic "fit" standpoint, and will enhance a company's competitive position. Table 2.4 summarizes key topics usually included in (but not limited to) a business case.

A note of caution: business cases usually have best case timelines and cost estimates. Many times, budgets, and schedules are assigned based on these estimates. However, these sometimes turn out to be unrealistic. Budgets and schedules should be refined with greater detail after formal approval during the project planning phase.

Table 2.4 Business Case Contents.

• Summary and Introduction	• Legal and regulatory issues
• Objectives	• Estimate costs and timeline
• Description of new product/project	• Resources needed
• Why should the company do this (what is driving this effort)?	• Social implications
	• Alternatives considered
• Cost-benefit analysis (ROI, NPV)	• Recommendations

Table developed by David Tennant

It is appropriate to note that investors and shareholders are generally risk averse. The business case is intended to provide a comfort level to decision makers; and to show that significant thought and research has been conducted to ensure the venture will be profitable. Companies that do not perform a business case for new ventures are taking unnecessary risks.

The Roles of Marketing and Engineering in Product Development

Someone from either department may serve as the project leader. However, the two groups generally have differing perspectives on priorities and timing which can lead to conflict.

Beyond the product, the marketing team is concerned with revenue, profitability, placement, and all of the previous topics discussed in this chapter. As a result, the marketing leader will be schedule-driven and concerned about containing costs. Remember, all of the costs of product development must be recaptured in the product's pricing. Therefore, this can translate into a perception of impatience by other departments.

Engineering on the other hand, is focused on designing, testing, and delivering a product that is error free (a definition of "quality" by the way). By training and education, technical professionals have a tendency to downplay fast-tracked schedules and to a lesser extent, budget adherence. To the engineering leader, having a flawless (or "perfect") design and manufacturing process is more important.

Consequently, the groups have differing perspectives and accountabilities, but are charged with coordinating their efforts to deliver a new product. Table 2.5 shows these differing priorities.

Regardless of which team is leading the development effort, a certain amount of tension can be beneficial in forward progress. There will always be "give and take" in the design process. Figure 2.6, Product Development Process Flow, shows a sample process flow in the development of new products. This can vary between companies, but it provides a general sense that the development of new products is not a random series of actions, but rather, has well-defined steps with checks and balances along the way.

Table 2.5 Marketing vs. Engineering Focus in New Product Development.

Marketing	Engineering
• Cost conscious in meeting budgets, ROI (Return on investment) • Schedule compression • Developing advertising campaign • Freezing design • Changes saved for next release (V2.0) • Anxious to beat competitors to market (first to market gets market share)	• Desire to do it right the first time • Changes to improve product are necessary for success • Schedule slips are acceptable with justification • Budget is important, but having a quality product is more important • Testing and Quality Control are paramount • Scope changes are part of the process

Table developed by David Tennant

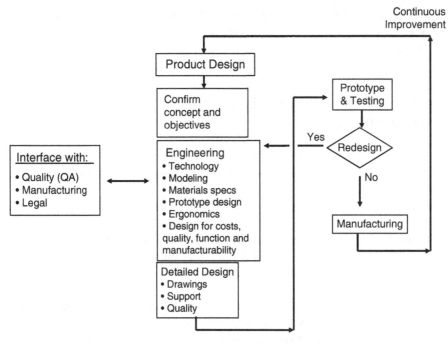

Figure 2.6 PD Process Flow. Credits: Diagram by David Tennant

The role of the product manager is to provide leadership to the project. This includes:

- Ensuring the team has the tools, training, and support needed to be successful
- Resolving conflict and building team morale
- Coordinating the efforts of various departments working on the project
- Serving as a conduit of information between groups
- Serving as a direct link between the project team, executive leadership and external suppliers
- Ensuring trust and a good working relationship between team members
- Being cognizant of project quality, budgets and schedules
- Ensuring all stakeholders are identified and key stakeholders are kept informed.

A product manager must be an active project participant and cannot ignore conflict, warning signs of failure, or the desires of key stakeholders. The role, whether from engineering or marketing (or any other department), is focused on managerial and leadership skills more than technical skills.

Marketing Services

Although services can be thought of as a product, they must be marketed differently. When you buy a new computer or smart phone, you can look at it, feel if, and hold it; perhaps even try it out to see how you like it. If you purchase the product and don't like it, you can usually return it for a refund or exchange.

Services are different. For example, if you provide accounting services to clients, it's an intangible service. You cannot feel, smell, see, taste, or try it out. You may or may not get a refund if you're unhappy with the result.

Like any product, one must obtain user input. What would a customer like to see in the way of customer service? Would guarantees of satisfaction make the product more desirable? What is the outcome if your client purchases the service (Benefit)?

Also, the use of data analytics can be useful. What are data analytics? Data analytics is a process to analyze raw data so that conclusions may be drawn from that information. In today's technical world, a lot of the data can be analyzed using automation and algorithms. Data analytics can help manufacturing identify and remove roadblocks. It can help firms become more efficient in their business processes.

However, for services, it can also help a business understand their customers, and therefore improve their marketing campaigns, personalizing of their services and improve their products. This data can be collected from historical records or from customers directly. For example, a utility can pull up data from their customers showing electric use patterns, payment options used, and time of day peaks in electric use. This data can be segmented. From this data, new rates (a service) can be derived and offered to their customers. Other sources of data include companies in the marketplace that collect and sell data. Think about your web browser capturing you internet trail. They know when you're most active (time of day), the types of sites you typically visit and what kinds of purchases you make online. For example, data analytics may be able to track and tell what music sites you visit, what topics you watch on YouTube (history videos for example), and what types of car accessories you like to purchase.

Services must have a different marketing strategy and data analytics can help a business hone their message, offer new services, help segment their market, and determine which online platforms would be most effective.

New Product Development and Market Economics: The Future of Electric Trucks vs Costs and Public Policy

In the United States, electric vehicles (EVs), account for 2% of new car sales (2021). However, that number is expected to rise to 30% or more in the next 10 years. It is commonly known that cars are evolving from the internal combustion engine (IC) to electric motors. EVs are slowly seeing greater acceptance in the marketplace. Along those same lines, large trucks are also moving in this direction, but the outlook isn't as clear.

This shift in transportation offers opportunity, rewards, and risks for companies looking to profit from this disruptive technology. This section will focus on EV trucks as a viable market opportunity.

Public Policy

A number of countries have mandated that the IC engine will no longer be sold in new vehicles by the year 2030. This includes several European and Asian countries. The United States, in comparison, has not mandated a ban on ICs, but has public policies that push the USA in this direction. For example, there are significant tax credits (savings) for individuals and corporations who purchase alternative fuel vehicles. At the same time, taxes on existing fuels (gasoline and diesel) have been rising (increasing operating costs).

Consequently, the market is steadily moving from traditional fossil fuels to cleaner, alternative-fueled vehicles. For large-rig Semi trucks, the waters are a little murkier. For instance, the long haulers would need a range of 400 miles and a network of charging stations along the way. However, how long will it take to charge an 18-wheeler? If more than 30 minutes, this could be a barrier to entry. Also, the trucking industry has also been looking at natural gas-powered vehicles and hydrogen powered vehicles.

From a new product development standpoint, the market seems to be a moving target. In the United States, there are eight basic classifications of trucks as follows:[7]

Classes 1 through 3: light, non-commercial trucks such as pickups, SUVs, and minivans

Classes 4 through 6: medium-sized commercial trucks.

- Class 4 – Gross vehicle weight is between 14,001 lb. to 16,000 lb. Super duty pickups and walk-in box trucks fit this description.
- Class 5 – Gross vehicle weight is between 16,001 lb. to 19,500 lb. This class includes larger commercial walk-in trucks, delivery trucks and bucket trucks.
- Class 6 – Gross vehicle weight is between 19,501 lb. to 26,000 lb. School buses, weighing will generally fit into this category. Class 6 is also where a Commercial Driver's License requirement appears. This also includes delivery trucks (UPS, Fed Ex, etc.).

Classes 7 and 8 are generally the big rig trucks, as follows:

- Class 7 – Gross vehicle weight is 26,001 lb. to 33,000 lb. These trucks are city street sweepers, garbage trucks, large buses, furniture trucks and smaller semi-trucks.
- Class 8 – generally "severe duty" trucks such as dump trucks, cement trucks and the large semis such as Peterbilt's, Freightliner's, and Kenworth's. Many of these trucks have three axles (or more) and weigh over 33,001 lb. Some of the larger rigs can weigh up to 80,000 lb.

All the above classes of vehicles have prototype or production models using batteries and electric motors in place of IC engines or diesel motors – even the large Freightliner trucks. However, the adoption and use of truck EVs is far from certain and economics plays a huge role in its future.

For instance, small trucks, pickups etc. are now available to the public in EV configuration. The Ford F-150 truck, one of the most popular trucks on the U.S. market, is now available using electric motors in place of the IC engine.

The determination of electric motors will be driven by vehicle usage combined with the initial cost and operational costs.

For example, Categories 1 through 3 and 4 through 6 trucks will be suitable for use during the day and charging at night. This is most likely a strong market for truck EV sales.

Delivery trucks such as those used by UPS and Fed Ex are good candidates as they are delivering during the day and parked (for re-charging) over night. School buses would be similar as they pick up and discharge children in the morning and late afternoon; then are parked overnight. The initial cost of electric vehicles is higher than traditional IC engines, so usage will play a large part. Table 2.6 show data on average annual miles driven.

If we consider the United States Postal Service (USPS), their full fleet of delivery trucks uses 85 million gallons of fuel per year.[3] The current average price for gasoline (July 2021) is $3.22, so annual fuel costs for the whole USPS delivery fleet would be:

$$85 \text{ million gallons} \times \$3.22 \text{ per gal} = \$273.7 \text{ million dollars annual fuel costs.}$$

One EV USPS delivery truck electricity cost: $5,100 \dfrac{\text{mi}}{\text{Yr}} \times 1 \dfrac{\text{kWh}}{\text{mi}} \times \dfrac{\$ \, 0.08}{\text{kWh}} = \$408 \, / \, \text{yr}.$

212,000 trucks × $408 = $86,496,000. This means, operationally, USPS could save about $187 million per year in fuel costs by switching to EVs. Of course, this does not consider the following:

Additional Costs

- Capital costs to install electric charging stations
- The *initial* higher cost of EV trucks

Cost Savings

- Federal tax credits for switching to EVs
- Significantly lower maintenance costs (primarily tires and brakes).

There are approximately 212,000 vehicles in the delivery fleet. The USPS fleet would be an optimal target for EV sales as they are out for delivery during the day and back to the fleet station to charge overnight.

Table 2.6 Truck Category Average Annual Miles.[2]

Category	Average Annual Miles Driven
Delivery truck (e.g., UPS)	12,500
School bus	11,000
City transit bus	42,000
Class 8 semi-truck	63,000
Postal Service (USPS) long-life trucks	5,100

Table 2.7 Potential New EV Opportunities.

Opportunity	Companies or Industries
Charging station manufacturers and installers	Siemens, General Electric, Tesla, and many others.
Truck stops and gas (charging) stations	All along highways and secondary roads, a new, extensive network of charging stations will be required. Current gas stations, malls, and restaurants are likely users.
Utilities	EV charging may represent large electric sales in off-peak (overnight) hours. This may increase the need for new (but cleaner), power plants.
Manufacturers and dealers	All car and truck manufacturers and their dealer networks will be on the forefront of EV technology implementation.
Battery manufacturers	Those firms that will supply auto/truck makers with new lithium or similar batteries.
Battery recycling	With more and larger batteries on the road, at some point there will be a need to recycle or safely dispose of car/truck batteries.
Rare mineral extraction companies (i.e., lithium)	Most EV car batteries use lithium as a core component. Lithium is not common in all countries and there are a limited number of companies in this specific mining extraction business.
Home builders and electrical contractors	EV will require car charging stations in both existing homes and new homes. This represents a lucrative business opportunity.

Table developed by David Tennant.

There are many opportunities for companies looking to affiliate with the coming EV market. Table 2.7, on New EV Opportunities, outlines a few of them.

Note that disruptive technologies can provide opportunities for new business creations in the areas of new products, product support, spin-off businesses, and jobs. Federal, state, and local areas consider subsidies as a positive feature for the following reasons:

- Subsidies are used to move the economy from one form of business to another. In the case of EVs, from a polluting energy fuel to a cleaner fuel source.
- It creates jobs, and therefore economic activity and a tax base.
- New technologies keep industry competitive at home and abroad.

The above table lists just a few of the opportunities that EV vehicles will offer to large companies and entrepreneurs alike. There will be many spin-off companies as a resulting shift in transportation focus.

Other Marketing Considerations

It must be emphasized that numbers on paper always look optimistic and tough questions should be asked in our market evaluation for both cars and trucks. Consider the following:

With both cars and trucks growing in popularity, this means the utility electric system, the "grid," will see additional loads. In the United States, the grid system is not entirely consistent. In the Southeast, there is ample excess power, so straining the system is not considered an issue for the short term. Longer term planning will certainly be needed.

Table 2.8 Utility Model EV costs.

Fleet Type	Number of Vehicles	Daily Miles Driven	Miles Driven per Month	Daily kWh Used	Monthly kWh Used	Electric Range (miles)	Charger Type	Battery Size (kWh)	Acceptance Rate of (kW)	Charging Time Frame	Number of Charging Hours (Empty to Full Charge)	Energy Consumption Rates (kWh/mile)	Average Power Demand (kW/hour)	Fleet's Cost-$/kWh	Fleet's Electric Cost per Month	Gas Cost $/Gallon	Miles/Gallon	Fuel Cost per Month
Light (Up to 8,500 lbs. GVW)	1	50	1500	20	600	338	Level 2	135	15	Overnight	9.0	0.4	2.22	$0.08	**$48**	$3.50	17.00	**$309**
Medium (8,501-26,000 lbs. GVW)	1	100	3000	100	3000	200	DC Fast Charger	200	120	Overnight	4.0	1	25.00	$0.08	**$240**	$3.50	10.00	**$1,050**
Heavy (Over 26,000 lbs. GVW)	1	125	3750	250	7500	300	DC fast Charger	600	100	Overnight, Mid-route	4.0	2	62.5	$0.08	**$600**	$3.50	8.00	**$1,641**

Spreadsheet courtesy of Cobb EMC.

Some areas of the country, primarily on the West Coast, utilities may struggle to meet this new demand without building new power plant capacity. How does this impact your marketing targets for EVs of any kind? Further, California has designated that only zero emission cars can be sold beginning in 2035. How will this impact the electrical system? How will this impact utilities, car dealers, and most of all, the residents of California? What does this mean if you are a car dealer in California? If you're an entrepreneur, does this open an opportunity?

Many utilities have been shifting from coal to cleaner forms of energy, primarily natural gas, solar and wind power. If the price of natural gas increases (while the price of gasoline might decrease), what are the implications? As a marketing professional, does this make you rethink the opportunities or do alternative opportunities arise?

Most likely, a marketing group would develop several models (that is, an Excel spreadsheet) that consider the implications for various increases or decreases in the price of gasoline vs. the price of electricity. To some extent, the regulatory issues are out of your control, but one would expect the impacted parties: utilities, car dealers, and the public would offer opinions to their elected officials.

In the Southeast, a number of utilities are employing various modeling techniques based on anticipated adoption of EVs (refer to Table 2.8). Some of the questions to consider include:

- Current EV use is 2% of the total car market. What strain, if any, does this produce on the electric system if that use rapidly accelerates to 10%? 20%? 30%?
- How much additional power would be needed in the next 10 to 20 years?
- If new electric generating stations are needed, what kind of fuel should be considered? Nuclear? Alternative (wind, solar), natural gas?
- It takes anywhere from two years (wind, solar), to five years (natural gas) to 14 years (nuclear) to build specific types of generation. When should we start planning for new construction?
- Where should these plants be located (usually near a water source)?
- In which direction is the regulatory wind blowing regarding fuel choice?
- What is the impact if rural areas do not adopt EVs, but large cities do?

Electric Vehicle Discussion Questions

1. As a manufacturer of new Semi trucks, where should your development efforts be focused? Electric? Natural gas? Hydrogen?
2. For electric trucks, this means that the national electric grid must be able to support millions of vehicles charging simultaneously overnight. Some areas of the country have plenty of power to support EVs; other areas are short and are struggling to provide power during normal and peak periods. How does this factor into your company's products?
3. Public policy can change. With each election cycle, changes mean that public policies regarding the environment can change. Is it possible have a voice or influence so that alternative vehicles are more fully supported through legislation?

Chapter Key Points

- Strategic planning is important as it charts a company's path to future profitability.
- Marketing and Sales are different disciplines. Marketing is concerned with the four Ps; sales is concerned with obtaining new customers and contracts.

- The four Ps include Product, Promotion, Pricing, and Profits.
- Product development includes four stages: Introduction, Growth, Maturity, Decline.
- It is common for companies to use spreadsheet models to determine costs, pricing, profitability, using sensitivity analysis.
- Breakeven point is where all development costs have been recovered and profitability of the new product starts.
- A business case analyses a new product's feasibility, potential in the marketplace, SWOT review, and other factors. This can also be called a business plan or feasibility study. Its results are used by senior management to make a decision (go or no-go).
- Marketing and Engineering sometimes have different goals and accountabilities. This can lead to organizational conflict.
- The role of Product Manager is a leadership position.

Discussion Questions

1. The project leader has an important role. While all of the below are important, which is the most important and why:
   a. Communications
   b. Integrity and trust
   c. Conflict resolution
2. Some companies believe that a Product Development Manager must have strong technical skills. Is this true? Why or why not?
3. What value does the Marketing Group bring to the product development effort?
4. If the cost of each new product is $40, and our company expects margins of 25%, should we price our product at $50? Why or why not?
5. One of the costs that may be in our product's development is legal costs. What kind of legal activities would be incurred?
6. Why is it important to know the break-even point?
7. What are methods to anticipate customer demand and pricing?
8. Projects should always exist to support the corporation's strategy. Why?
9. Once products have reached the decline stage in the product lifecycle, they become a commodity. Is this true? What is a commodity?
10. All costs in a product's development should be recovered in the first year of a product's sale in the marketplace. Is this a good strategy? Why or why not?

Answers to Case 2.1 Discussion Questions – Kodak

1. As a large, successful US company, Kodak was once a "dogs of the Dow" firm listed on this stock exchange. It could be argued that Kodak's management was unwilling to change or consider new ideas to displace or augment its line of film products. What does this tell us about the role of innovation in a company?
   This case should reinforce the need for companies to continuously develop new products. This could also be an effort to update existing products.
2. With the advantage of twenty-twenty hindsight, it is easy to criticize Kodak's lack of foresight or imagination. However, it is likely that past success blinded Kodak to the future of film. This might be categorized as a resistance to change, which is found in most large corporations; it becomes part of the corporate culture.
   If you are in a senior level position – at any company – how would you work to change the corporate culture?

Changing the corporate culture is extremely difficult from middle or even senior management levels within a company. Corporate culture describes "Who we are" as a company. Generally, when the Board of Directors recognize a company is stagnating, they usually bring in a new CEO to turn things around or go in a different direction. Even with new leadership, it will still be difficult to move change through the company's ranks. This is why new CEOs will many times clean house at the senior levels and bring in people they can trust to implement change.

It is unlikely anyone below the CEO can implement change; and most executives with any powers of observation will see deterioration of a company's position, and several will probably leave long before a new CEO takes over.

3. What role should the marketing department have played in this long-term decline?
The role of marketing is to be aware of industry trends, provide input to the corporations strategic planning, and generate new ideas and products to keep the company in front of its competitors.

4. From a managerial standpoint, what are the challenges when a company is in decline?
Once a company is on a downward spiral, it presents serious managerial issues:

- *There is a rush to maintain revenues, which means more pressure on salespeople.*
- *R&D is pressured to devise new products for the market – and quickly. This leads to mistakes or the wrong products at the wrong time.*
- *Once the decline is noticed by employees, the best people – those who have the skills to land a new job fast – are the first ones out the door. The firm is then left with the "B Team" to perform a company turnaround.*
- *A publicly held company, such as Kodak, will face mounting pressure from analysts and shareholders to "do something."*
- *Publications like the Wall Street Journal and others will start to cover the firm's issues, putting more pressure on management.*
- *If the senior leadership of the team cannot think more creatively, the Board of Directors may replace the CEO (and other executives) with new leadership.*
- *From a marketing perspective, customer needs and wants should be determined quickly and action taken. It should have been noticed by the marketing team much earlier that film was going to become obsolete with digital cameras. Would senior management have listened?*

Answers to Discussion Case 2.2 – Portable Battery Unit

	Sale Price			
10% of 16.9 million	Price Sensitivity	TTL Revenue	Margin	Margin %
1,960,000	$ 50.00	$ 98,000,000.00	$ 24,500,000.00	25%
1,960,000	$ 45.00	$ 88,200,000.00	$ 19,404,000.00	22%
1,960,000	$ 40.00	$ 78,400,000.00	$ 15,680,000.00	20%
1,960,000	$ 35.00	$ 68,600,000.00	$ 11,662,000.00	17%

1. What are the advantages and disadvantages of this kind of modeling?

   It is acceptable to build models that can look at different scenarios. While the model above is somewhat simplistic, most companies use very sophisticated models to predict sales, margins, etc. The advantages include evaluating different market acceptance scenarios and its impact on revenues and margins.

   Disadvantages include the possibility that market acceptance may be way off target, impacting costs and revenues. Further, the true costs of the product should be as accurate as possible to obtain realistic outcomes.

2. Why do you think margins per unit decline as the price drops?

   There could be several reasons for this.

   - *Efficiencies of production ($/unit) tend to be beneficial at higher levels of product manufacturing. These efficiencies typically drop as the units produced decrease.*
   - *Most likely, suppliers of raw materials provide volume discounts, so less raw material ordered will result in higher material costs.*
   - *The fixed costs of production (salaries, overheads, etc.) are being absorbed by fewer units of production.*

3. Looking at the sale prices compared to margins (%), is $50 the best price? How do you know students will pay this amount for your product?

   In this case, we do not know that $50 is the best price. It will provide the most profitability if we can sell almost two million units, but there is no guarantee for that level of market demand. Also, consider their target market, students, whom are very price sensitive. The best way to determine the selling price is to conduct focus groups with the target market.

4. If 10% of students purchase your product the best-case scenario shows $98 million in revenue and profits of $24.5 million. Worst case is profit margin of $11.662 million. In your opinion, are these numbers realistic?

   If those sales goals are obtained, the numbers are a good approximation. However, this cannot be considered absolute. What if we find some areas of the country will pay $50, but other areas only $30? This model will most likely need to be very dynamic with frequent updates on sales, costs, and revenues. And perhaps parsed to address different market acceptance across geographic areas or income levels.

5. Do you think you will have competitors prior to reaching the maturity stage?

   If the product "takes off" it will only be a matter of time before competitors appear. This may happen prior to the maturity stage but is hard to predict. It will depend, to some degree, on how technically difficult the product is to copy.

6. Is this type of analysis realistic?

   Model building to approximate sales and revenue is very common. However, the flaw in this model is the assumption that 10% of the total student population will buy the new product. Some considerations:

   - *If only 1% of the student population accepts the product, it could be a money loser*
   - *If 20% accept, we would perceive this as a good thing. However, at 20% acceptance, this means manufacturing would need to produce twice as many units to meet demand: 4 million units instead of 2 million. Can manufacturing meet this? Can we get the raw materials at the price and time needed? Can we distribute that much product in a timely and efficient manner?*

This type of analysis is very common, but fraught with potential errors. As mentioned previously, focus groups can provide valuable insights and help prevent the above dilemmas. First, focus groups will provide a more realistic answer to what students would actually be willing to pay for the proposed product. Further, this will also provide some insights as to whether students think this is a product they really need (or want) and can use. Students are very price sensitive, so any product competing for their dollars must offer value (or the perception of value).

To make assumptions of market acceptance based on total student population and a blanket target rate is very risky and is probably unrealistic.

7. What role would the engineering or R&D groups play in this evaluation?
   The marketing department would be the most qualified to develop pricing, distribution, and margin scenarios (remember the 4 Ps). However, based on marketing research and studies, the results can have a direct impact on R&D and engineering. Some items to consider include:

   - *If students are only willing to pay $30 per unit, can a cheaper model be produced? R&D will look for alternative technologies or develop new ones. It may also include reducing some of the features of the unit in order to reduce costs/price.*
   - *Engineering will work closely with marketing and R&D. Can less expensive materials be used, but still provide a quality product? Can we consider how the units are produced for more efficient methods? Can batch production be used rather than a continuous run? Can we find a supplier (or multiple suppliers) that can provide similar raw materials at lower cost?*

8. The table above shows gross revenues and net profits. There is no mention of the cost of the product. What are some of the costs that would be involved? Can you determine from the chart above the total "costs" of the product?
   The chart does not provide any indication of product costs. Some of the costs to be included in the product's price include:

   - *Cost of raw materials*
   - *R&D and engineering design costs*
   - *Manufacturing costs: fixed and variable*
   - *Overheads: salaries of employees, benefits, legal department, sales group, etc.*
   - *Costs of distribution (transport of product to stores)*
   - *Marketing and PR (advertising)*
   - *Intellectual property (royalty payments?).*

   If these costs can be spread over a large number of units, the cost per unit will be low and the margin per unit will be higher. Spreading cost recovery over a longer time frame will also be beneficial (say three years instead of one).

Answers to Discussion Questions

1. The project leader has an important role. While all of the below are important, which is the most important and why:
   a. Communications
   b. Integrity and trust
   c. Conflict resolution?

While each answer is important, the correct answer is b. A project or product manager is a leadership role, and one of the principal hallmarks of leadership is trust. This will be of high importance in leading a product development team to success.

2. Some companies believe that a Product Development Manager must have strong technical skills. Is this true? Why or why not?

   The Product Manager will need leadership and managerial expertise more that technical expertise. The PM should expect to have technical subject matter experts (SME) assigned to the team. The required PMs skills include vision and focus on the objectives, strong communications and relationship building up and down the organization, managerial excellence, ability to track budgets and schedules, and the ability to resolve conflict. There are additional skills that could be listed, but these are the fundamentals.

3. What value does the Marketing Group bring to the product development effort?

   - *Marketing may assign a Product Manager to lead the development effort*
   - *Marketing tracks industry trends and determines the 4 Ps*
   - *Marketing will develop pricing/margin scenario spreadsheets*
   - *Marketing will determine the break-even point*
   - *It is likely that marketing will work with the finance group to determine NPV and other financial metrics*

4. If the <u>cost</u> of each new product is $40, and our company expects margins of 25%, should we price our product at $50? Why or why not?

   At a cost of $40 and a selling price of $50, the required 25% margin will be realized. However, there is no guarantee that $50 will be accepted by the market. If the price, for example, must be dropped to $45 to sell, then the required margin will not be met. The company would then need to determine if a 20% margin is acceptable or find ways to cut costs so that the required profitability can be obtained.

5. One of the costs that may be in our product's development is legal costs. What kind of legal activities would be in our budget?

   The legal department, depending on the product, may have a very key role to play in product development. Some of the activities may include:

   - *Developing and reviewing contracts with suppliers*
   - *Submitting patents for the new product(s)*
   - *Determining intellectual property ownership and if royalties will be paid*
   - *Registering trademarks or copyrights*
   - *Outlining contracts or guidelines for product distribution rights*
   - *Advising the product team and senior management of potential legal (product liability) or regulatory issues.*

6. Why is it important to know the break-even point?

   Breakeven is related to profitability and NPV (net present value). It is important to know the breakeven point (the point where all costs are recovered, and the new product is profitable):

   - *If the breakeven point is too long (say 10 years), this may not be acceptable and prove to be a money-losing product.*

- *A long breakeven point will give competitors time to catch up and almost guarantee that your product will never meet margins.*
- *Each company is unique and will determine what is an acceptable breakeven based on time, NPV, costs, and price.*

7. What are methods to anticipate customer demand and pricing?

- *Customer focus groups.*
- *Brand and reputation of company*
  Note that brand recognition can offer the perception of quality, thereby allowing a higher selling price.
- *Evaluating market acceptance based on similar products from competitors.*
- *Studying history of previously released products by your company. This may provide <u>insight</u> into pricing and market acceptance.*

8. Projects should always exist to support the corporation's strategy. Why?
The company develops a strategy, or long-term plan, to remain profitable. The corporation issues new products that capitalize on its strengths. Consequently, all activities: projects, new product development, etc. should support the corporation's strategy. If a new product is not part of the strategic plan, why should it be funded or supported?

9. Once products have reached the decline stage in the product lifecycle, they become a commodity. Is this true? What is a commodity?
In the traditional sense, commodities are defined as base or raw materials. This includes items such as oil, wheat, sugar, gold, etc. These are considered primary commodities. However, secondary commodities include products that are produced using primary commodities, such as gasoline and plastics from oil, cereal and bread from wheat, etc. Commodities, unprocessed, are placed into three categories: agriculture, energy, and metals.

To a lesser degree, products once they reach their maximum potential may be called a commodity, but this is not in the strict definition. This basically means that a product in "commodity" status is simply not able to command a high price and must compete with many other producers.

10. All costs in a product's development should be recovered in the first year of a product's sale in the marketplace. Is this a good strategy? Why or why not?
This is an unrealistic expectation. The goal, of course, is to recover all costs and meet the breakeven point as quickly as possible. For more sophisticated products – e.g., newly developed cars (Tesla, for example), sophisticated electronics, and specialized applications (medical equipment) – the breakeven point will be longer. It is up to each company to determine an appropriate time frame to recover costs and become profitable.

It is safe to assume that most companies will not expect cost recovery in the first year.

References

1 Innolytics.AG https://innolytics-innovation.com/what-is-innovation.

2 Fortune 500, Full List 1980, https://archive.fortune.com/magazines/fortune/fortune500_archive/full/1980.

3 Fortune 500, Full List 2020, https://fortune.com/fortune500/2020/search.

4 Ty Haqqi, Yahoo Finance, 3 August 2020, https://finance.yahoo.com/news/10-most-profitable-companies-america-215951022.html.

5 Photosecrets.com, viewed 20 August 2020; https://www.photosecrets.com/the-rise-and-fall-of-kodak.

6 Yahoo Finance, viewed 20 August 2021, https://finance.yahoo.com/quote/KODK?p=KODK.

7 Fullbay, Making Sense of Truck Classification, viewed 3 August 2021, https://www.fullbay.com/blog/truck-classification.

Bibliography

Aaker, D. (1998). *Strategic Marketing Management*, 5th e. John Wiley and Sons.

Belz, A. (2011). *Product Development*. McGraw-Hill.

Cooper, R. (2011). *Winning at New Products*. New York: Basic Books.

Grewal, D. and Levy, M. (2018). *Marketing*, 6th e. New York, NY: McGraw-Hill Education.

Hubspot Blog post. https://blog.hubspot.com/marketing/product-life-cycle. (viewed 10 August 2021).

Norman, D. (2013). *The Design of Everyday Things*. New York: Basic Books.

Patel, N. https://neilpatel.com/blog/product-life-cycle (viewed 15 August 2021).

Patterson, G. (2012). *Million Dollar Blind Spots*. Issaquah, WA: AudioInk Publishing.

3

The Role of the Engineering Group in Product Development

First, it is important to recognize what a product development engineer does. Typically, the product engineer must have a four-year degree in engineering. The role can encompass different types of engineering. Initially, when manufacturing was a huge part of the economy, it would most likely be a mechanical or production engineer developing new products. Today, that role has expanded to include industrial engineers, computer engineers, electrical engineers, and many similar technical disciplines.

Think about your everyday products such as tools, electronics such as computers, smart TVs, and your smartphone. Larger products include your car (and its components: transmission, engine, body), the plane you fly in (Boeing, Airbus, Cessna, Piper, etc.), and common household products such as cleaning supplies, clothing, and even the food you eat are harvested or processed using machines designed by engineers.

Driving Products – the Engineering Perspective

In product development, there are a number of stakeholders that must be involved in the project. Typically, Marketing or Engineering will lead (or even co-lead) the project. Engineering is focused on designing, testing, and delivering a product. Engineering, by training, experience, and education will have a high focus on developing a high-quality product. What is quality?

There are many definitions:

> Some describe it as simply meeting customer needs or expectations. Others describe Quality as "freedom from defects."

The International Standards Organization (ISO) defines quality as:

> The totality of characteristics of an entity that bears upon its ability to satisfy stated and implied needs.[1]
>
> – ISO 9000

> A product or service free from deficiencies.[2]
>
> – American Society for Quality

Product Development: An Engineer's Guide to Business Considerations, Real-World Product Testing, and Launch, First Edition. David V. Tennant.
© 2022 John Wiley & Sons, Inc. Published 2022 by John Wiley & Sons, Inc.

The engineering team, as a matter of professional diligence, a "can-do" attitude, and pride of accomplishment, will always be concerned that the design is appropriate and of high quality. Engineers do not like to see their designs fail, or products that they have put much thought and creativity into become associated with a failed product. It is for this reason the engineering profession is held in high esteem by society. Science and engineering have been responsible for some of the greatest technical achievements in history: landing men on the moon, development of the internet, satellite communications, and functional infrastructure such as roads, bridges, water systems, the electrical grid, etc.

However, for purposes of product development, we need to think of quality as having two distinct flavors: Product Quality and Managerial Quality. What does this mean? Fundamentally, a company cannot have success unless it has both technical and managerial excellence (quality). How good will the new product be if we have miscommunications, overrun the budget and schedule (moving toward financial failure), and fail to do robust testing during the product's development? It is desired that the design, which is a form of technical planning, and project planning, which is a form of managerial planning, have high levels of quality. For managerial quality, this means robust teamwork, exceptional leadership, keeping stakeholders informed, etc. Project Managerial excellence, a key component of success, will be discussed further in Chapter 6, Product Initiation.

Engineering will include quality using the Quality Assurance (QA) and Quality Control (QC) processes. What is the difference?

QA is the overall quality program that will serve to confirm that quality processes and procedures are being followed. It is the overall umbrella for quality activities. This is the development of processes and audit procedures. QA will provide confidence that the product (or project) will meet relevant quality standards. This is generally considered a managerial function.

Some of the tools that can be utilized for QA include:

- Benchmarking: Compare current products with other, successful products
- Flowcharting: System flowcharts that illustrate how system elements are related
- Checklists
- Design of experiments to statistically determine the most desirable characteristics at a reasonable cost.

QC is a subset of QA and is the actual inspection and testing of the product during production. An example of this would be inspecting parts on the assembly line and removing those that do not meet specifications. Essentially, QC monitors the results to determine if they comply with relevant quality standards and identifying ways to eliminate causes of unsatisfactory performance, even going so far as to stop the production line until the root cause of the problem has been determined. QC is generally considered a technical function.

A few tools that can be used by QC include:

- Inspection
- Variable analysis
- Sampling
- Statistical probability techniques
- Control charts

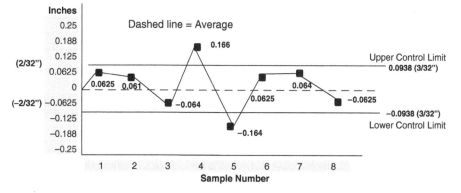

Figure 3.1 Control Chart. Figure by David Tennant

There are always costs associated with a strong QA/QC program. Some of these include training, testing, audits, and controls (procedures and processes).

However, the costs of not having a robust QA/QC program can be much more expensive and may include scrapped parts, rework, warranty work (fixing defective products at company expense), product recalls, product liability lawsuits and damaged company reputation. This will lead to loss of revenue.

A control chart, such as that shown in Figure 3.1, indicates an upper and lower control limit. If the component being inspected is between these limits, it is acceptable, beyond these limits, it must be rejected.

It is important to note that customers will pay more for a product that is high in quality or even has the perception of quality. Consider, for instance, the luxury brands of automobiles. Some brands command very high prices for their reputation (whether deserved or not) of a quality product. Consequently, many companies promote their products in meeting quality or customer satisfaction ratings based on consumer research groups such as J.D. Powers™ and Consumer Reports™. Therefore, there is a strong link between Marketing and Quality.

In summary, quality must take a high priority in developing a successful product. For large or small companies, the lack of quality can lead to bankruptcy.

Engineering Disciplines

It is likely that several different engineering disciplines will be used in developing new products. Table 3.1 Description of Engineering Disciplines, lists the types of engineering areas that are in various fields of manufacturing, software, energy, aviation, transportation, infrastructure, and product development.

Why is the distinction of engineering disciplines in the above table important? Primarily because on any project or new product development, there will be a variety of technical disciplines engaged. These disciplines will depend on the type of product under development.

For example, if we are developing a new wind turbine application, the following engineering talent would be involved:

Mechanical engineering: turbine gearing, materials of construction, fasteners, blade design, unit size, and location (best wind conditions).

Table 3.1 Description of Engineering Disciplines.

Engineering Discipline	Industries Served
Aeronautical engineering	National defense, general aviation, space programs
Industrial engineering	Any environment in which production occurs; primarily concerned with efficiency of production, manufacturing flow, time and motion studies.
Computer engineering	Computer engineers are focused on writing software code, testing, troubleshooting, developing new software, interfacing between devices and organizations. The last time you used your ATM card, computer algorithms and software ensured your card worked flawlessly.
Mechanical engineering	Generally involved with any type of mechanical product (engines, transmissions, etc.) and can be found in all industries including utilities, aerospace, manufacturing, biomedical, etc.
Electrical engineering	Critical for the design of circuits in all products requiring electrical current: cars, airplanes, medical devices, computers, electronics, buildings, power plants and the national power grid (electrical transmission and distribution).
Chemical engineering	Chemical engineers are focused on processes to produce chemicals, pharmaceuticals, biological products, food processing, alternative and primary fuels, adhesives, etc. It is desired to produce these products on an industrial scale in manufacturing facilities.
Production engineering	In any manufacturing facility, you will find production engineers. For example, in a light aircraft (or car) assembly plant, production engineers are concerned that the machines used to produce parts, seats, windshields, body panels, etc. are the correct machines, properly programmed, and running efficiently. They are always looking for ways to improve product manufacturing. For example, think of a soft drink can. This required set up with the proper machines, running at the right speed, and moving from one station to another without breakdowns or quality variation.
Biomedical engineering	Focused on using engineering principles to improved healthcare and associated devices (medical imaging, orthopedic implants, pacemakers, etc.).
Civil engineering	Civil engineers focus on a broad array of applications including structures (supporting building columns, beams, etc.), dams, roads, and bridges.
Instrumentation and Controls engineering	I&C is a relatively new engineering specialty now recognized for its own applications. This includes factory automation, control of processes (power plants, chemical plants, water treatment, etc.), and building automation. Most processes are automated and computer controlled, which means that pumps, motors, compressors, turbines, and other large machinery are programmed and controlled with sensors and controllers from a software-based system. I&C engineers play a major role in specifying the computer controls and sensors (instrumentation) that will control the manufacturing process.
Additional engineering disciplines:	Nuclear, Packaging, Ocean/Marine, Petroleum, Mining, Agricultural, Health and Safety, and Environmental.

Table developed by David Tennant.

Instrumentation and Controls engineering: Determines safe operating limits, how and when to start/stop the turbine, and interface with other turbines on the grid. This group will specify, procure, and program the needed instruments and software for safe and efficient operation.

Civil engineering: Design of foundation for wind turbine base and supporting structure. The support structure (most likely a pedestal) must be able to withstand strong winds exhibiting a force on the turbine in addition to the turbine movement and vibration.

Electrical engineering: Design of electrical components in the current generating part of the turbine (essentially, the electrical generator) and how the turbine will interconnect with the electrical grid. Also, most wind turbine "farms" may have up to 100 wind turbines. It is important how each is interconnected to the grid and that safety features are designed so that one turbine failure does not trip (i.e., cause other wind turbines to fail) the whole wind farm off the grid. They will develop power and wiring diagrams.

It should be apparent that in any new product development effort, a variety of engineering talent will be utilized. So, who will lead the product development effort, Marketing or Engineering? This will depend on the company's organizational structure and the type of product in development. However, the engineering leader may lead the overall project, or just the engineering piece. Regardless, this person will be in a key leadership role and needs general or project management skills to be successful. Either discipline may lead the new product effort, engineering will play a key role in bringing the product to fruition. So, what exactly will the engineering team do?

The Engineering Process

Figure 3.2 Engineering Design Process is an approach engineers use to solve problems. A new product usually has technical challenges such as implementing new standards, complying with federal or state regulations, performing research, designing, and testing the new product, and implementing solutions based on test results and customer feedback.

The engineering team will follow a common approach to identifying the product challenges and solving the technical problems. The common approach is illustrated in Figure 3.2 Engineering Design Process.

Assuming the engineering leader has been involved with the development of the business case and strategic planning at the senior level, it is time for he or she to begin the design process.

Here is a brief description of each step of Figure 3.2.

Define the Problem

The strategic planning of the firm, combined with a detailed business case should outline the new product's features, requirements, target market, etc. It is now up to the design team to develop a product that meets those requirements. As demonstrated earlier, a design team will most likely have a multi-disciplined approach, using different types of engineers and designers.

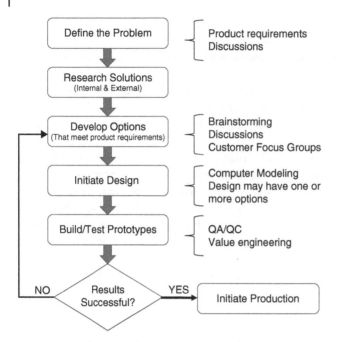

Figure 3.2 Engineering Design Process. Figure by David Tennant

The new product's development should also have a tentative timeline and cost estimate. The design team must determine if the product can be developed within the budget and schedule provided. Some of the questions to be addressed in this stage include:

1. Based on the requirements, what materials are best suited (plastics, metal, composites, etc.)?
2. Which vendors should we approach for the outsourced materials?
3. Can we build the new product, with the required features, in a compact package weighing less than _____ pounds?
4. Can we produce this product within the budget and timeline?
5. Which stakeholders should we approach for clarification?
6. What customer preferences do we know? What is the target market(s)?

Most of the questions posed at this stage should be answered by the business case, the marketing team, the executive sponsor, and key stakeholders. Clearly if any end user preferences exist (i.e., from focus groups), this is highly desirable information.

Research Solutions

There is always the possibility that elements of the new product may have been researched or developed by someone in the company or an external source. Nonetheless, a first place to start is always internally. However, in our quest to develop the new product, here are several considerations for research:

- If a new technology or manufacturing process is needed, universities are always doing research. Reach out to the chair of the engineering department for suggestions and referrals.

- If your company has an R&D group, can they assist? Have they already been involved?
- There are many specialists and consulting firms that have the potential to assist.
- Can you lure away key people from your competitors to join your team?
- There are many technical societies where building relationships can assist in finding expertise. For example, The Institute of Electrical and Electronics Engineers (IEEE) is a large technical society with a very diverse and skilled membership.
- An internet search will always reveal someone, somewhere who is working on a similar problem or obstacle.

This approach should give the design team an idea of what materials, circuits, robotics, or other requirements will be utilized. An engineering team leader will be responsible for coordinating the various disciplines and keeping the team focused on the primary objectives.

During the research phase, it is important for the engineering team to remain aware of the target market, price points, and meeting customer needs. These should have been identified early on in any market analysis. However, it is worth noting that in the middle of solving design issues, it is easy to forget these objectives and start designing a new product for ease of manufacturing instead of ease of use for the customer. This means the design team must understand the commercial and business issues, the target market, and functional requirements as the design progresses. It never hurts to ask for clarification.

Develop Options

Through the research and discovery process, it will become apparent that several approaches may ultimately provide the solutions to the design questions. It then becomes necessary to select the option that best meets the products requirements at a best value. Notice best price was not mentioned, but rather best value. It is easy to cut corners or expenses, but will it help in developing a high-quality product? Best value is a preferred approach: high quality for a reasonable price and also considering the expense of warranty work or loss of company reputation when releasing an inferior product.

It is appropriate at this point to confer with the manufacturing experts (production and industrial engineers) to determine the costs and manufacturability of each option.

- Will new capital manufacturing equipment (e.g., numerical control machines or robotics) be required?
- Have we considered outsourcing some of our manufacturing for this product? Where or who would be qualified to produce our new product. (Author's note: There are a number of firms that align companies with specialized manufacturers.)
- The most promising design options should be shared with the product development team for a consensus decision.
- What will be the cost per unit when producing 10,000 units, 50,000 units, or 100,000 units? That is, will economies of scale be present? (Most likely, yes).

Initiate Design
The new product's design will require coordination between the engineering disciplines and manufacturing. This coordination can occur with weekly engineering

design meetings, status reports, and project meetings with the leadership of the product development team.

It is at this point those preliminary drawings are now put into motion based on the selected design option. This will require detailed analysis and possibly modeling to choose the optimum design. By optimum design, this means the new product design that will meet quality standards, provide customer satisfaction, is safe to operate, and cost effective to manufacture.

Design usually includes calculations, modeling, drawings using a computer aided design program (CAD), checks for accuracy, quality, etc., and issue of design drawings for manufacturing. It may also be necessary to build one or a few prototypes.

A prototype is an actual first-run product item produced in the factory. It is meant to be evaluated, tested, and subjected to stress and misuse.

The results of this testing will be used to determine if changes are needed before the product is manufactured in large, batch runs. It should be noted that modeling of the new product does not eliminate the need for a prototype(s). It will, however, refine the design allowing fewer prototypes to be produced; and at the same time speed up the design process. Figure 3.3 shows a design analysis of a wind turbine blade.

The analysis can show stress points under different operating conditions thereby allowing engineers to refine the design to shore up weak points or allow for material substitutions. In another application.

Figure 3.4 Engineering Modeling of Bicycle Frame shows the stresses on a bike frame based on the design shown. Several iterations of the design can be performed, thereby allowing the engineer to select the most valuable design. The criteria may include costs, strength of materials, expected lifespan of the frame, reliability of the product and safety. Suppose the frame is designed for a bike rider that weighs 180 lb. What if someone weighing 200 lb. rides it? 250 lb.? Would it fail?

The model allows engineers the opportunity to model "What if?" scenarios and provide a robust and safe design.

This means that several prototypes could be produced and tested, based on the model. It is easy to see how this type of modeling can save time; and improve product quality and safety. Further discussion on modeling is provided in Chapter 8.

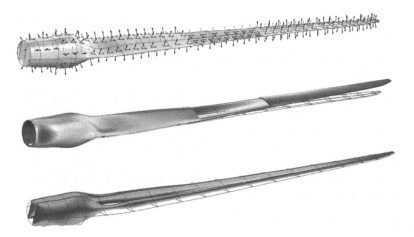

Figure 3.3   Engineering Model – Wind Turbine Blade. *Source*: Image made using the COMSOL Multiphysics® software and is provided courtesy of COMSOLL

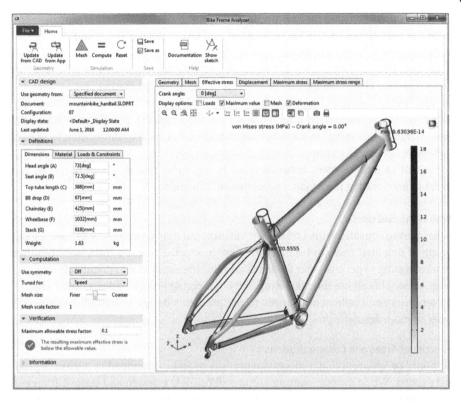

Figure 3.4 Engineering Model – Bicycle Frame. *Source*: Image made using the COMSOL Multiphysics® software and is provided courtesy of COMSOL

Build and Test Prototype

If we consider the bike frame in Figure 3.4, the model will allow the company to build one, two, three, or more prototypes based on the computer simulation. When assembled into a fully functional bike (it is possible the wheels and shift mechanisms were also modeled), it can be tested with a variety of riders and ride applications: sport, mountain biking, casual, etc. and with riders of differing height, weight, and body shape. It will become apparent from test data which of the three prototypes exhibits the best ride characteristics: handling strength, stability, safety, costs, longevity, etc. It is possible the model may have to be redesigned, refined, or changed based on the test data, but the modeling can provide a short cut to reaching the optimum design.

Note that good engineering practice dictates that there is always an appropriate Safety Factor (S.F.) applied. The bike may be intended for people weighing 200 lb., but will be designed, for example, with a 25% safety factor, meaning it will support a weight of 250 pounds. A factor of safety is built into most products: ladders, elevators, speed ratings on auto tires, etc. This does not mean one is to push the labeled limit but is meant as a cushion for those who may misuse the product or unknowingly overload it.

Initiate Production

Once the final design has been tested and approved, a set of manufacturing drawings will be issued as "Final, Ready for Production," meaning the manufacture of the new product is to begin using the approved drawings. Recall that the manufacturing plant built the prototype and will begin production upon drawing receipt and the "green" light from the new product team.

Recall Table 2.5 in Chapter 2, Marketing vs. Engineering Focus in New Product Development. There will usually be tension between the marketing and engineering groups. By nature, and training, engineers tend to be very analytical and detail-oriented, with a desire to issue a high-quality product. This is a priority over schedules and budgets. Marketing, on the other hand, prefers to beat the competition to market (schedule fast track), minimize design changes, and develop at low cost. Consequently, there will be disagreement between these two groups. This is where modeling can assist in accelerating the product development schedule.

Who Is the End User?

A key part to consider is the end user or customer. It may be necessary to have a focus group of potential users to beta test the new product. There are instances of limited market testing to get customer feedback and impressions prior to manufacturing on a large scale. This allows the marketing and engineering groups to determine customer preferences and implement changes to the product's design or manufacturing for a safer or more desirable product.

Functional Areas and Communications

It should be apparent that a successful product must be developed by a team of product experts. While the engineering group will fulfill the design and testing, it takes a full complement of skills from many functional areas to ensure success. As stated in the previous chapter, the Marketing group will play a large role in the conceptual vision of the product and may indeed lead the project. Finance and accounting play equally important roles to ensure the product is profitable. Supply chain will assist in procuring products, services, or consulting talent necessary for the product's success.

Sales will need to understand the product to obtain customer orders. Additional stakeholders will include company executives, department heads, and key suppliers.

In essence, each functional area is important to the success of the new product and, consequently, communications and cooperation across functional departments needs to be open and honest. It will take a special leader to drive the project. A number of these topics will be presented in subsequent chapters. Further discussion on communication techniques will be provided in Chapter 5, Moving Forward.

What Is a Project Engineer?

The terms project manager and project engineer are sometimes used interchangeably. However, there is a difference. A project manager generally has responsibility over the complete project, reporting to the executive sponsor and overseeing the efforts of all functional areas (marketing, engineering, R&D, production, etc.). A project engineer, on the other hand, will lead the engineering/design team. His or her responsibilities are confined to the technical issues and development of the product. While in a leadership role, the position will report to the project (or product) manager.

Ergonomics (Human Factors Engineering)

Have you ever rented a car and had trouble finding the gas hatch release? Did the dashboard displays initially appear confusing and hard to understand? And the numbers or letters on a dark background are not easily seen. Welcome to products that are poorly engineered from a human-user perspective.

What is ergonomics? In essence, it is the design of products that are user-friendly and will comfortably fit 90% of the population, both male and female. The term "human factors" is also commonly used, often interchangeably, with "ergonomics."

The study of anthropometrics deals with the size of various populations. For example, one can find data that describes the smallest female (5% of population) to the largest male (95% of population) and design a product that fits those groups from 5% to 95%. If your company manufactures restaurant dining chairs, there is enough data available so that your chairs will easily fit 90% of the population. Appropriate designs can also be provided for drivers reaching the steering wheel, or the standardization of buttons and symbols on your TV remote controller. Consider the following:

- Some people, especially those that are tall or of large frame, find it difficult to enter and exit some sports cars. This is because some of these cars are very narrow and low to the ground. From this, can we infer that some sports cars are made to appeal to younger drivers that are shorter and slim?
- A common problem is the difficulty older people have opening medicine bottles. This is due many times to arthritis or loss of hand strength with age. Some companies have made their products more useable with better (but still safe) packaging. Others have ignored this problem or not recognized their product is not user-friendly.
- Occasionally, when entering a store for the first time, do we pull or push to open the door? Unless it is clearly marked, we have only a 50% chance of being right.

If you have a TV remote control device, you may have noticed that the symbols, and often the button placements, are identical from one remote to another, regardless of brand. For example, looking at Figure 3.5 Typical TV Remote, it is easy to understand the keys as these same symbols are used almost universally. Can you identify the play and pause buttons? The power button?

How did the science of human factors begin? It was noticed during World War 2 that some planes were having very high accident rates due to pilot error, not combat. As humans, we all have capabilities and limitations, but normal operations are different from emergency operations. It was found that some controls and displays were confusing and that pilots were manipulating the flaps when they had meant to manipulate the landing gear (or vice versa). Even with many hours of flight training, this was a problem. It was determined that control placements were fine for routine missions. However, once stressful or emergency conditions occurred, pilots were unable to relate to their controls. As a result of extensive research, airplane controls became standardized, much the same way cars have standard displays and controls. For example, you will find almost all cars have the transmission shifter on the floor between the front seats. It would not make sense to locate the shifter on the door panel or on top of the dash. Similarly, your dash display is directly in front of you to enhance your ability

Figure 3.5 Typical TV Remote. Erik-Mclean / Unsplash

to drive while noticing your speed with a glance. In many ways, these controls and displays have been standardized. How does this come about?

It is common for industries to have technical working groups that develop standards that all manufacturers agree to use or comply with. For example, standards are developed for the car industry by the Society of Automotive Engineers. Pressure piping and fasteners are developed by the American Society of Mechanical Engineers, etc.

Notice in Figure 3.6 Speedometer Display, that it is easy to determine the speed with the digits in whole numbers (10, 20, etc.). If our speedometer measures in odd units, as shown on the right, it is harder to tell at a quick glance our true speed.

We find our perception plays a role in our expectations ... or some would say conditioning. We are used to seeing whole digits and research suggest that we more easily "read" this type of gage. The author has had some experience with driving foreign cars in Europe where speedometers have different conventions than what we are used to in North America.

What are the implications for controls and displays in modern jet aircraft (military or civilian)? Controls in a spaceship? A nuclear power plant? All these examples are much more complicated than an automobile. Misreading a dial or gage in a jet aircraft could have fatal consequences. This is why aircraft displays have some commonality. This commonality is not by coincidence, but by design. What if our pilot suffers from fatigue or lack of sleep? Can we design our displays to be more easily read and understood even when the pilot is sleep deprived?

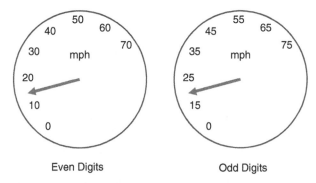

Even Digits   Odd Digits

Figure 3.6   Speedometer Display. Figure by David Tennant

A True Scenario

On the morning of 28 March 1979, there was an accident at TMI, the Three Mile Island nuclear power plant located in Harrisburg, PA. At around 4 am, cooling water started to escape through an open valve in one of the two reactors at the plant site. For over two hours plant operators failed to correctly read the symptoms and failed to close the valve. They also mistakenly shut off an emergency cooling system that would have operated satisfactorily if left in automatic mode.

In the control room, dozens of alarms were ringing, and the situation quickly spiraled out of control. While no major release of radioactivity occurred, the top of the fuel bundle was uncovered (water levels dropped during the accident) and significant damage to the plant reactor occurred.

Consequently, it was determined that operator error compounded the problem and that there were approximately 100 additional nuclear control rooms around the country with the same basic design philosophy and layout. It was further determined that the control room layout was not conducive to safe operations and that poor human factors was an issue. It wasn't so much the error of operators, but rather the way controls, panels, and displays were arranged that made a major accident a high probability; especially in emergency or high stress scenarios.

As a result, robust operator training – for both normal and emergency conditions – was implemented for all plant operators in the US. And all nuclear plant control rooms were evaluated from a "human factor" or man-machine interface perspective. Corrections were made (at great costs) to ensure all power plant control rooms would be user-friendly and easier to use when diagnosing problems, which would lead to appropriate corrective actions.

The author participated in a national BWR (Boiling Water Reactor) owner's group and surveyed multiple nuclear control rooms around the country. Some of the findings violated common perspectives or expectations:

- Some control knobs when turned to the right decreased the parameter on the display. This could be pump flow, water level in a tank, etc. An analogy might be when we turn the volume knob on our home stereo to the right, we expect the volume to increase. Having the volume decrease would violate our expectations or conditioning.

- Other problems with nuclear power plants included controls or displays that were too high or low on a panel. Or the units on a gage were not easily understood: there were instances where displays showed odd numbers rather than even (consider our speedometer example).
- Gage indicators that should have had a range of 0 to 30 psi might have a range that indicates 0 to 300 psi, which means our maximum pressure of 30 psi would barely register on this gage's display. This is a case where the units displayed did not correlate with the required "sensitivity" of the item being measured.
- In many instances, labeling of components was too wordy or redundant, making quick analysis of a situation time-consuming and challenging.
- In all nuclear plants, drills were performed to observe operator reactions. This is called task analysis. It was found that during normal operations, the operators easily manipulated controls and understood next actions. However, during emergency scenarios – that is, high stress events – the operators were sometimes confused, slow to determine the root problems and made errors in taking corrective action.
- It was found that color coding was not consistent and sometimes did not correspond to industry guidelines.

So, what does this have to do with new product development? Why is this true story important? The designers of nuclear plant control rooms, up to that time, had designed the plants from a logical or engineering perspective, not from an operational point of view: this did not help the plant operators (end users). If your company is designing new products, take heed of the lessons from Three Mile Island or other poorly designed products:

- Design the product with the end-user in mind.
- Design it for potential misuse whether intentional or accidental.
- Expect your product to be used in unexpected ways.
- Design your product to be safe whether used in normal or under high stress conditions.
- Do significant product testing including deliberate misuse.
- The customer/end-user should be able to reach all controls (e.g., in a car, plane, or boat) and see the associated displays without expending extraordinary effort.
- If the user interface is a computer screen, use industry guidelines that include standardized symbols and color coding (some people may be color blind).
- Do not violate expected user conventions (right for up, left for down, etc.).

Note that the use of appropriate human factors standards in your product design will help prevent accidents or misuse of your product. Human factors is a design philosophy that enhances safety and efficiency. It also allows the design of products to deal with the psychology of people and their expectations. It can help your firm reduce the possibility of product liability lawsuits.

Additional Design Considerations – Product Liability

A woman ordered hot coffee from a popular fast-food chain. After leaving the drive-through window, she stopped to add milk and sugar. The coffee spilled in her lap and scalded her. She sued the restaurant and won a sizeable settlement. Who was at fault?

Could this have been prevented? This is a true story and is discussed more fully in Case 3.1 later in this chapter.

A Fictional Scenario

At her home office, Sheila works at her desk and sits in an office chair that has swivel wheels; the chair can also swivel around its base by 360 degrees – in other words, a typical office chair that can be purchased at any office supply store. The chair allows her to move from her desk to a printer behind her without getting out of her chair. She needs to retrieve a book from the top shelf of her bookcase. Not wanting to go to the garage for a ladder, she wheels her office chair to the bookcase and stands on it to reach the book. Unfortunately, she loses her balance from the chair and falls to the ground hurting her back. She files a lawsuit seeking medical and punitive damages. Her case could be argued that, although she used the chair for a different than intended purpose, this should have been foreseen by the manufacturer.

The company would argue a defense of "product misuse" by the customer. However, consumer protection laws indicate that if the misuse was reasonably foreseeable, the manufacturer (and possibly the seller) may not escape liability unless measures were taken to prevent against harm resulting from misuse.

The above examples illustrate accidental use or misuse of a product. Could these incidents have been prevented? Is it the company's responsibility to "fool proof" products so they cannot be misused? The answer to these questions is "Yes" based on product liability laws. There are many other examples of manufacturers being sued for the misuse of their products. Therefore, other aspects that must be considered during the design phase of our product is:

- What is the intended purpose of the product?
- Will people be capable of misinterpreting the product's use or purpose?
- Can we prevent people from intentionally or accidentally misusing the product?
- What is the "experience" that people will gain from the product?
- Have we considered ergonomics in our design?
- Have we done adequate consumer beta testing to ensure ease of use and safety?
- Have we considered the likelihood of product liability potential (i.e., a legal review)?
- Do we have adequate safeguards in our product?

Companies do their best to launch products that are safe and high in quality. However, mistakes occur, and customers can be injured using or misusing a product. There is a perception that many frivolous lawsuits are filed; however, these tend to be minimal in number. There are legitimate instances of poor product design or cheap construction, and the courts are the remedy used by consumers for restitution. Companies genuinely attempt to remedy product injuries as they do not want the risk of damaging their brand identity, reputation, or losing customers, all of which lead to declining revenues.

It is possible to anticipate the misuse of products. If your product is not clearly labeled, has unclear instructions, or does not have a safety interlock system (e.g., the computer control system in your car), it is likely that the product will be misunderstood and possibly misused.

It should also be noted that products manufactured in other countries may not follow US or European safety standards. Therefore, the seller or importer, should determine the imported product's suitability for the market, especially for the US and EU markets.

Government Oversight – Consumer Protection in the United States

There are a number of Federal agencies in place to investigate defective products and provide consumer protections. Below are a few examples.

Labeling and Packaging

There are both State and Federal guidelines in product labeling and packaging. Essentially, product labels must be accurate and use language that is commonly understood by the public. This can include products such as foods (content), tobacco and alcohol products, fire hazards (For example, clothing, bedding, and other fabrics made with non-fire-resistant materials), and product use (Ladders, power equipment, etc.).

Figure 3.7 Ladder Labels, shows the use of warning labels on ladders meant for personal use. Does this labeling relieve manufacturers of liability?

Figure 3.7   Ladder Warning Labels. Photo by David Tennant

Food and Drug Administration[3]

The FDA is charged with protecting the public health by assuring that foods are safe, wholesome, sanitary, and properly labeled; ensuring that human and veterinary drugs, vaccines, prescription over-the-counter medicines, and medical devices intended for human use are safe and effective:

- Protecting the public from electronic product radiation
- Assuring cosmetics and dietary supplements are safe and properly labeled
- Regulating tobacco products
- Advancing the public health by helping to speed product innovations.

The FDA oversees medical devices (implants, hearing aids, pacemakers, etc.). Any manufacturer wishing to sell a medical device must first apply to the FDA for approval. They must present evidence that the device is reasonably safe and effective for a particular use.

Product Safety

The Consumer Product Safety Act (1972) established the Consumer Product Safety Commission with extensive power regarding consumer safety. The CPSC has the power to ban products it deems unfit or unsafe for consumers, keeps a database of faulty products, investigates claim by consumers of faulty products, and does testing on products. The agency has removed products in the past that dealt with unsafe toys, baby products, and items containing asbestos.

Environmental Protection Agency (EPA)

There are both state and federal laws pertaining to environmental protection. The mission of the federal EPA is to protect human health and the environment. Should your manufacturing require an addition or further enhancements, it is likely the state and/ or federal EPA will be involved. Will the increase in production require more power? Will large amounts of water be required? Will the facility discharge water to streams or rivers? Will air emissions increase?

In the United States, each state has an equivalent EPA to manage state permits and ensure compliance with federal regulations. Each state has unique circumstances. For example, Florida has the Everglades to manage and protect. Each state regulates public and private facilities for air quality, water quality, hazardous waste, water supply and discharge, solid waste, surface mining, underground storage tanks, and in some situations, state parks or coastal areas.

Occupational Safety and Health Administration

It should be apparent that companies developing new products must be aware of the various statutes and regulations that are directly applicable, and also to any facility changes made to accommodate the new product's production. An example is OSHA, the Occupational Safety and Health Administration. While not directly applicable to products, it does apply to the manufacturing facilities where the new product may be produced. Essentially, OSHA is concerned with companies providing a safe workplace for production workers. Therefore, the new factory your company may build to support the new product will be subject to compliance with OSHA regulations.

There are other agencies that deal with deceptive advertising, internet, telemarketing sales, credit protection, and environmental protection, etc.

Legally, a product liability case is where a product (or service) failed and caused injury to the user. Let's consider further the case details of the woman who spilled hot coffee in her lap.

Discussion Case 3.1 – Lawsuit over Hot Coffee[4]

On February 27, 1992, a 79-year-old widow was in the passenger seat of her Ford Probe ordering a Meal at the drive-through window of an Albuquerque, New Mexico, McDonald's. Since there were no cup holders in the Probe and the interior surfaces were sloped, her grandson, the driver, pulled into a parking spot after they received their order.

"I wanted to take the top off the coffee to put cream and sugar in," Liebeck told a local news station at the time. "So, I put the cup between my knees to steady it to get the top off."

The coffee spilled on Liebeck's lap, resulting in second- and third-degree burns over 16% of her body. She went into shock and was hospitalized for a week, undergoing numerous skin graft operations.

When Liebeck's medical bills topped $10,000, she contacted McDonald's and asked to be reimbursed. But McDonald's responded with an offer of $800.

At the time, McDonald's required its franchises to brew its coffee at 195–205 ^{0} and sell it at 180–190 ^{0}, far warmer than the coffee made by most home coffee-brewing machines. McDonald's reps suggested that the blame lay with Liebeck for holding the cup between her legs.

However, the trial revealed that Liebeck was not alone. McDonald's had received more than 700 complaints about burns from hot beverages over the previous ten-year period. After seven days of testimony and four hours of deliberation, the jurors sided with Liebeck. They awarded her $200,000 in compensatory damages. But, because she had caused the spill, they reduced the amount to $160,000. The jurors then awarded her $2.7 million in punitive damages, which, they reasoned, was equivalent to about two days' worth of McDonald's coffee sales. The total was $2,735,000 more than Liebeck's lawsuit had requested. The full award was reduced later to $500,000, but still much higher than the original requested compensation.

Discussion Case 3.1 Questions

1. There could be many parties at fault, including the victim. Besides the elderly woman and McDonald's, who else in this scenario could be a lawsuit target?
2. Is this a legitimate lawsuit? Why or why not?
3. Was McDonald's at fault or the widow? On what do you base your conclusion?

4. Do you think this lawsuit triggered additional lawsuits against McDonald's or similar restaurants?
5. If you oversaw quality control or product testing at McDonald's, what would you do now, post-verdict?
6. Has the threat of product liability lawsuits made products (and services) safer?
7. Would more effective labeling on the cup provide liability protection?

Here are a few high-visibility product liability cases:[5,6]

- Monsanto has dealt with numerous legal challenges claiming their product, Roundup, causes cancer. Although Monsanto was later acquired by Bayer company (a German-based firm), Bayer not only bought the assets but is also dealing with Monsanto's liabilities: in this case, product liability lawsuits.
- Dow Chemical company and Corning, Inc. reached a settlement for $3.2 billion to settle a class-action lawsuit by customers claiming their breast implant product caused bodily damage and injury. The company also agreed to pay women who wished to have their implants removed.
- Asbestos made by Owens-Corning was alleged to cause cancer and premature death. The company settled a class-action lawsuit agreeing to pay $1.2 billion. Ultimately, the number of claims increased to the point the company was unable to pay and resulted in bankruptcy.
- Johnson and Johnson Talcum Powder Product liability. In June 2020, Johnson and Johnson agreed to a $2.1 billion settlement to women who suffered from ovarian cancer due to use of this product. JNJ no longer sells talcum powder in the United States but does in overseas markets.

It should be apparent that intended product use and misuse must be considered in any new product development. Manufacturers should evaluate this potential during the design and "beta tests" for misuse of their products. Failure to consider legal repercussions can be financially devastating.

Design Challenges – Product Misuse

People are unpredictable and some will misuse a product. The design team will need to explore different ways the product can be misused (even intentionally) and design safeguards to prevent this. Research (i.e., a beta test) can illustrate the different ways a customer can misuse the product. Some products have built in safeguards. For example, a table saw has a blade guard and auto cut-off switch to prevent or minimize accidents.

Design the product for ease of use or for a great experience. If your product is complicated to use or understand, this will increase the likelihood of misuse.

In the case of medical devices, how likely is a doctor to make a mistake when sleep deprived? Can we anticipate mistakes? Or a tree cutter using a power saw in a dangerous way to save time? Can we anticipate the mistakes or injuries that could occur?

Figure 3.8, Product Misuse Review, illustrates a review process for anticipating intentional or accidental misuse.

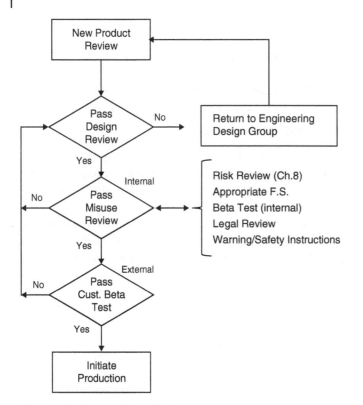

Figure 3.8 Product Misuse Review.

It is impossible to plan for every potential misuse of a product, but a reasonable effort should be made to mitigate or prevent obvious misuse. The use of a product risk assessment, covered in Chapter 8, can assist in this effort.

Intellectual Property is a consideration for ground-breaking technology or a new innovative product.

Does a patent already exist for the process or product you are developing? This means time and money to research patents. If none, you will need to budget time and money to apply for your company's patent protection. If a patent already exists, then your team will need to determine how to make your new product unique to avoid a patent infringement lawsuit.

There is a tendency to jump on the "trends" bandwagon. While consumers indeed want the latest smartphone technology or computer games, do not follow the crowd. One will always be behind the curve with this strategy.

The author had the opportunity to work with a large, multi-national equipment company headquartered in the upper mid-west USA. The company made good products, but instead of innovating new equipment, they simply looked to mimic what their competitors were doing. Consequently, they were not considered an industry leader.

The proper strategy is not to imitate your competitors, but leapfrog over them to become the industry leader. Market research and new technology can assist in accomplishing this objective.

Do not design products for today's market, design for the market five years in the future. Both engineering and marketing need to collaborate with such a strategy.

Problems with Product Development

1. Lack of market research on customer requirements, needs, and wants. It is easy to design to *our* tastes and preferences, but is this what the customer wants? As stated previously, there are many avenues for companies to explore customer requirements: focus groups, surveys, research (competitor products), and data collection (analytics).
2. Being a market follower instead of a market leader. The best companies will always be in front of market trends, or designing for tomorrow's market. Being a market follower will always put you, at best, in second place. Design products for the future and leap over your competitors.
3. Staying in your core competency. It is necessary that companies must change and adapt to stay in business. However, the company's strengths should be used to develop new products or services, not its weaknesses. This means that a car company doesn't suddenly decide to start producing smart phones instead.
4. Tradeoffs of costs vs. quality. There will always be ways to cut costs; however, quality is a general concept with which the buying public is aware. There are countless stories where great ideas failed because too much emphasis was on low costs. There have been many times where metal gears were replaced with plastic ones to save money. Of course, plastic gears under heavy use will fail. A strong company brand and reputation are key to profitability. Poor quality will tarnish your reputation. And warranty work may erase any profits from your new device.
5. Inefficient manufacturing can lead to higher production costs, poor workflows, and ultimately a lot of scrapped parts or rework. A worst-case scenario is product recalls or high levels of warranty work. In today's competitive environment, efficiency in manufacturing is required. This implies reliable production machinery, well-trained machine operators, streamlined factory workflow, and a quality control and assurance program for continuous improvement.
6. Lack of project management skills by team leaders. To take a new product from inception through to launch requires leaders who know how to put together budgets, schedules, motivate teams, and plan activities. Also, we can anticipate problems with risk reviews and strong communications. It takes both a special leader and one that has PM skills to be successful.

Chapter Key Points

- Many disciplines and functional areas are used in the development of new products.
- Engineers use a universal approach to solve problems and research ideas.
- The engineering team is responsible for developing a product that meets the market requirements.
- Engineers need to be creative and flexible in meeting market requirements.
- The customer or end user's needs and wants must always be considered by the product development team.
- The marketing team should share whatever customer research they have collected with the engineering team.
- There are many external sources to research product solutions: universities, technical societies, government labs, and consultants.

- Computer simulations of new products can save time and money during the design phase. It does not replace the need for a prototype.
- Testing of the prototype may reveal weaknesses or failure points. This will allow the engineers to revise the design.
- The quality of your new product must be designed into the product, not inspected. Quality is everyone's business.
- Human factors engineering should be a part of your product's review and development.
- To minimize the chances of product liability lawsuits, the design team must consider product abuse and misuse during the design and testing phases of the new product.

Discussion Questions

1. Reference Table 3.1 Description of Engineering Disciplines. Which engineering talents would be used to develop the following products:
   - A high-speed, commercial bullet train to offer express service at speeds of 200 mph between Chicago, IL and Dallas, TX?
   - A new electric car?
   - A new system that allows customers to order their groceries on-line?
2. What is a functional area?
3. Consider the fictional case where the user fell off her chair and hurt her back. What could be done to design and build a chair that prevents this action?
4. What is the role of engineering in developing new products?
5. What is the purpose of performing engineering modeling?
6. With engineering modeling available, why is a prototype needed?
7. There is always a debate about designing a product for ease of manufacturing vs. ease of use by the customer. List the pros and cons of each.
8. To produce a large quantity of a new product, your company will need to add capacity. This includes adding a new manufacturing line, placement of new capital equipment, and hiring more skilled workers. In addition, more electrical power and water will be required. What are some of the federal or state agencies that might be involved?
9. Both Ford and GM have announced plans to shift production away from internal combustion cars to electric vehicles. This has profound implications for how cars are designed and manufactured. Make a list of conditions or activities that each must consider in this transition.

Discussion Case 3.1 Questions and Answers – Lawsuit Over Hot Coffee

1. There could be many parties at fault, including the victim. Besides the elderly woman and McDonald's, who else in this scenario could be a lawsuit target? *Potentially, there are other companies that could be pulled into this lawsuit:*

   - *What about the automaker, Ford? Why would they build a car without cupholders when many Americans drive while drinking coffee or soft drinks?*
   - *The cup manufacturer. We have all observed lids that were difficult to remove or replace from fast food eateries. Why is it so hard to have a screw-on or clip-on lid, for example, for convenience (and safety)?*
   - *The maker of the coffee machine. Surely a manufacturer of commercial-grade coffee makers should be aware of the safe serving temperature.*

2. Is this a legitimate lawsuit? Why or why not?

   The victim in this case did most likely what millions of drivers do every day: pulled over to add cream or sugar to their drink. McDonald's had received previous complaints about the hot temperature of their beverage, which indicates they were aware of the problem yet did little to resolve it. Therefore, the author believes this incident was a legitimate case.

3. Was McDonald's at fault or the widow? On what do you base your conclusion?

   If the reader does some additional research on this case, it will be apparent that the victim's award of $2.7 million award was reduced. This means that the judge did recognize that some of the blame resided with the widow. However, the punitive damages against McDonald's were still significant as the company had known of the problem and ignored it.

4. Do you think this lawsuit triggered similar lawsuits against other McDonald's franchises or fast-food restaurants?

   Similar lawsuits for overly hot beverages were also filed against Burger King, Starbucks, and Continental Airlines.

5. If you oversaw engineering or product testing at McDonald's, what would you do now, post-verdict?

   There are several actions that can be taken because of this lawsuit:

   - *Develop procedures or guidelines pertaining to all hot drinks (i.e., tea, coffee, hot chocolate, etc.) and hot foods to ensure they are served at an appropriate temperature.*

   - *Set up a hotline or customer complaint resolution partnership with each McDonald's franchisee. This will bring trends or problems to the attention of the company's headquarters.*

   - *Be cognizant of customer complaints and potential lawsuits so that amends or settlements can be made before going to court (assuming legitimate events).*

   - *Have the Quality department (QA/QC) take a more active role. If a quality group does not exist, then this expertise should be developed.*

6. Has the threat of product liability lawsuits made products (and services) safer?

   Yes, products have had a larger focus on safety and prevention of misuse. While costing companies more money in design and manufacture, the costs of lawsuits can be much greater.

   Some companies have moved offshore to avoid US product liability laws; however, this does not relieve them of responsibility. The loss of company reputation, whether foreign or domestic, can have significant financial repercussions.

7. Would more effective labeling on the cup provide liability protection?

   Most coffee cups at fast-food chains have had "Caution, hot liquid" warnings on their cups for many years. Is it enough? Is it easily visible? It is likely that more warnings or visibility may shield a company from liability but would have to be resolved in a court challenge.

Discussion Questions

1. Reference Table 3.1 Description of Engineering Disciplines. Which engineering talents would be used to develop the following products:

   - A high-speed, commercial bullet train to offer express service at speeds of 200 mph between Chicago, IL and Dallas, TX.

- *Mechanical (train design, cabin design, modeling for stress and heavy loading)*
- *Electrical (electrical systems on the train)*
- *Instrumentation and controls (ensure safe switching between tracks, signal coordination in traveling through towns, ability to track train progress, obstacle avoidance, etc.), and controlling the train's speed and direction.*
- *Civil (design track routing, design of support structures or bridges train must cross over, water drainage).*

- A new electric car.
  - *Mechanical engineers – body, transmission, wheels/axles, brake systems*
  - *Electrical engineers – design of car electrical systems, battery*
  - *Materials engineers – material selection, development*
  - *Instrumentation – Design of car instruments and driver feedback systems*
  - *Manufacturing and Industrial engineers – ensure efficient factory workflow, order, and install capital equipment for manufacturing.*

- A new system that allows customers to order their groceries online.
  - *Software engineers and programmers to develop computer code.*

2. What is a functional area?

   - *A functional area is an area of expertise that resides as a department or silo in a company; for example, marketing, sales, engineering, accounting, etc.*

3. Consider the fictional case where the end user fell off her chair and hurt her back. What could be done to design and build a chair that prevents this action?

   - *Place clear labels on the chair (not very aesthetic)*
   - *Warnings and instructions in the user manual on how to safely use the chair*
   - *Design a locking mechanism so that the chair can be changed from rocking and turning mode to a stable, locked mode.*

4. What is the role of engineering in developing new products?
   Engineering is charged with the design of a new product:

   - *Using recognized engineering practices for a safe, effective design*
   - *Developing product specifications*
   - *Testing the new product along with test data collection*
   - *Preparing "Approved for Production" drawings*
   - *Troubleshooting and revising product specifications or design*
   - *Providing design continuous improvement*
   - *Work closely with Manufacturing to resolve production issues.*

5. What is the purpose of performing engineering modeling?
   Engineering modeling allows the engineer to perform "What if?" scenarios with material stresses, heat transfer, electromagnetic flow, or other parameters. The model will ultimately allow the engineer to design the optimal product. Modeling is useful in speeding up the design process.

6. With engineering modeling available, why is a prototype needed?
   Modeling allows engineers to create an optimal design. However, a prototype is necessary to confirm the design through testing and data collection. Based on the tests' results, the design will be confirmed as optimal, or may be changed. It may be necessary to have more than one prototype.

7. There is always a debate when designing a product about ease of manufacturing vs. ease of use by the customer. List the pros and cons of each.

   Ease of Manufacturing Pros: cheaper, easy for workers to build, many parts "off the shelf" from suppliers, quicker to market. Ease of manufacturing may not mean ease of use for the customer.

   Ease of Manufacturing Cons: cheaper may lead to product recalls and warranty work. Easily made means competitors can easily duplicate; poor quality may result in damage to the brand and the company's reputation.

   Ease of customer use Pros: builds customer loyalty, can market as user friendly and high quality for a higher price. May provide a competitive advantage, may reduce product misuse and thereby liability.

   Ease of customer use Cons: May cost more to design, test, and produce. Will require more thought and testing along with design changes (adding to costs and schedule). May delay product's early introduction to the market.

8. To produce a large quantity of a new product, your company will need to add additional capacity. This includes adding a new manufacturing line, placement of new capital equipment, and hiring more skilled workers. In addition, more electrical power and water will be required. What are some of the federal or state agencies that might be involved?

   New building addition will require compliance with building codes, national electric and fire codes, the Americans with Disabilities Act, OSHA (Occupational Safety and Health Administration), and, if there are discharges from the plant, possible state and federal environmental regulations (EPA). If the facility already has permits for water and/or air discharges, it may simply require revising existing permits.

   The company's insurance carrier may also have safety and operational standards or guidelines.

9. Both Ford and GM have announced plans to shift production away from internal combustion engine cars to electric vehicles. This has implications for how cars are designed and manufactured. Make a list of conditions or activities that each company must consider in this transition.

   - *Much of the existing capital equipment will not be needed. New production equipment may need to take its place. Disposal or reuse of the existing equipment should be considered.*
   - *The workflow will be revised. Raw materials will go through a different process. A new, efficient workflow will be required.*
   - *New materials will be needed, and suppliers must be secured.*
   - *Since electric cars have different characteristics to IC (internal combustion) cars, design will need to focus on transitioning customers to new controls, gages and displays (ergonomics).*
   - *The transmission will require redesign.*
   - *The power source, a battery, will need to be placed into the car and connected to the motor, transmission, and axles. The design will optimize placement and performance.*

Notes

1 ISO9000 Made Easy. https://iso9001madeeasy.wordpress.com/2018/09/15/what-is-quality-according-to-iso.

2 American Society for Quality. https://asq.org/quality-resources/quality-glossary/q.

3 Food and Drug Administration, What Does the FDA Do? https://www.fda.gov/ about-fda/fda-basics/what-does-fda-do.

4 Andy Simmons, Reader's Digest, 23 February 2021, "Remember the Hot Coffee Lawsuit? It Changed the Way McDonald's Heats Coffee Forever." https://www.rd.com/ article/hot-coffee-lawsuit.

5 Staff Author, The Five Largest Product Liability Cases, 26 April 2021, https://www. investopedia.com/the-5-largest-u-s-product-liability-cases-4773418.

6 Cooper & Friedman, PLLC, 8 December 2020, "Top Product Liability Cases in 2020," https://www.cooperandfriedman.com/top-product-liability-cases-in-2020.

Bibliography

Brickley, J., Smith, C., and Zimmerman, J. (1997). *Managerial Economics and Organizational Architecture*. McGraw-Hill.

Lawrence, A. and Weber, J. (2008). *Business & Society: Stakeholders, Ethics, Public Policy*, 12th e. New York, NY: McGraw-Hill.

4

The Core Team and Teamwork in Product Development

The Executive's Role in Product Development

It has been the author's experience that executives take their participation in product development or other projects very seriously. After all, they are financially accountable for the ultimate success (or failure) of the new venture, product, or project. But what exactly should their role be? At what point are they over controlling? Or, not engaged enough?

What exactly is an executive sponsor or project executive? In large organizations, an executive is typically understood to mean someone at the Director or Vice President level. It might also extend to the C-level executive, such as the CFO (Chief Financial Officer), the COO (Chief Operating Officer) or other C-level executive. In small and medium-sized companies, an executive might be someone at Senior Manager or General Manager level.

The fundamental point, however, is that senior management is held accountable for results, including new product development, launch, and profitability.

Working Within the Strategic Plan

As noted, every company has (or should have) a strategic plan. This can be a long-term roadmap to develop new products or services, a scenario to change the corporate culture, or a strategy to enter new markets or partnerships. The role of the executive in each of these scenarios is:

- Provide funding and support – In addition to funding, the correct resources, in the form of people, hardware, office space, software, and other tools, must be provided by the supporting executive. Projects cannot be successful without provision of the right people, at the right time, and for the correct duration. Since executives assign resources, this is also one way to prioritize products (projects). Products/projects that receive full funding and the best people, will be those that have the highest level of immediate support (i.e., the highest priority).
- Serve as a link between the product development team and the senior management team – The product development team needs the executive sponsor to support their

Product Development: An Engineer's Guide to Business Considerations, Real-World Product Testing, and Launch, First Edition. David V. Tennant.
© 2022 John Wiley & Sons, Inc. Published 2022 by John Wiley & Sons, Inc.

requirements at the executive level of the corporation. The sponsoring executive serves in this role and transmits information from C-level decision makers to the product team.

- Participates in the Gate Review process – On a regular basis, the current product's development should have a gate review. There is no formal defining point when this should occur but it should be decided by the executive and product team in advance as to when reviews will occur. It is conceivable that six, eight, or ten gates could occur during a product's development cycle. The purpose of a Gate Review is:
  - Determine if the product's development is on schedule and within budget.
  - Does it still fit within the company's strategic vision?
  - Is the product still relevant in the marketplace?
  - Does the proposed product still meet the financial objectives?
  - Have there been any significant changes or risks identified since the product's development began or since the previous gate review?
  - Are the initial drivers or market forces still present?
  - Can the Product Development Manager provide a convincing story that the product is worthy of continued support (justification)?
  - Provide funding to reach the next gate or cancel the development effort, if necessary.
- Ensure the team has the right tools, training, funding, and people to be successful.
- Provide "firepower" to resolve conflicts that the product team cannot resolve on their own. There will be times that executive support will be needed to resolve a conflict or remove an obstacle that the product team is unable to resolve. Primarily, it is possible the product leader is not at a high enough position to have influence with higher level managers, or they are having issues with a key supplier. It is in circumstances like these where the executive can assist in resolving the conflict.

  However, note that the team should exhaust all avenues to resolve an issue before escalating to an executive. Part of the PM's role is to manage and direct the product's development. When help is needed, the appropriate executive should be contacted.

  If you need senior management support to resolve an issue, always provide possible solutions. It is never a good idea to just drop a problem in the executive's lap. By presenting the executive with options, you allow him or her to choose one of your solutions or develop an alternative. If you present the problem without any potential solutions, you will be seen as simply a manager and not a leader; you will be perceived as lacking creativity, the ability to discern issues, or organizational awareness. One source of potential solutions is your product/project team – talk to them. It is always useful to offer several potential solutions.
- Confirm that new products and projects support the corporation's mission and strategic plan.

Project Management Processes

Generally, PM processes or procedures will not be implemented and followed without executive support. This means the executive team must deliver a strong commitment – perhaps even going so far as to mandate those processes must be followed. This can also be combined with compliance in all PMs annual reviews. This subscribes to the view that "What gets measured, gets done."

It is generally not difficult to obtain executive support: the executive team is responsible for implementing products/projects that support and achieve the company's strategic business objectives. Some companies have found that managing many projects simultaneously can be a challenge. The author is aware of one large Fortune 50 company that has upwards of 800 projects ongoing simultaneously at any time. One way that companies have addressed this complexity is to implement a PMO (Project Management Office). This is not suitable for all companies as it depends on the organization's size and the complexity of projects. Fundamentally, a PMO is put together to support or manage a company's projects. The PMO has several potential roles:

- To develop the managerial processes needed for product or project management.
- To provide support (e.g., by developing or tracking budgets and schedules).
- To bring consistency to product/project management across the company.
- To coordinate communications across projects and departments.
- To perform project reviews to determine if PM processes are being used.
- To develop best practices.
- To provide continuous improvement to PM processes.
- To directly manage products and projects as appropriate.

It is outside the scope of this book to discuss PMO development and implementation, but the reader should recognize the value a PMO can provide. If your company has one, the product development team should utilize its resources and practices in reaching its objectives.

Project management processes essentially offer structure and support to enable the product or project team to function efficiently. For example, how will scope changes be managed? It should be noted that changes in scope are common to many projects, but how these are managed is critical. Most companies will have a *scope change request* form that can be used to evaluate the value of the change. An example of a Scope Change Request (SCR) form is at the end of this chapter. Other useful tools may include:

- A standard format for status reports
- An action item log
- A stakeholder identification table
- A roles and responsibilities chart
- Risk review processes
- Flow charts for engineering design, supply chain, communications, etc.

Who Should Be Involved in Product Development?

Recall from Chapter 1, Figure 4.1 Core Team Product Development, in which a diverse range of groups can be involved with product development.

This is a reasonable model for a typical development team for the development of a product. However, we could easily add more stakeholders if needed. This is, after all, a model.

How do we, as product development leaders, determine who should be on our primary team and our extended team?

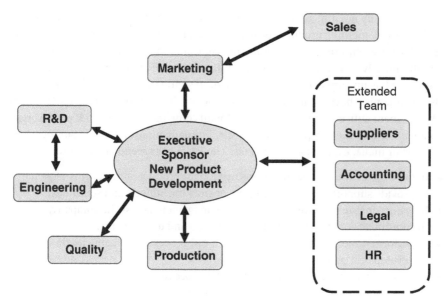

Figure 4.1 Core Team – Product Development. Figure by David Tennant

The answer is: "It depends." It depends on the complexity of the product under development, the extent to which each functional area is involved (including suppliers), and which executive sponsor is supporting your efforts.

The answer to this question will become more apparent as we drive through subsequent chapters. However, for our purposes here, it is appropriate to note that a core team will need to be in place to drive development. Let's consider a hypothetical product:

Hypothetical Product: A New Portable Printer

While there are many portable printers in the marketplace, your firm feels the time is right for a new generation of portable printer that is a new leap in front of its competitors. A feasibility study and focus groups have confirmed this proposition. Based on this research, and other factors, you have been charged with developing a new groundbreaking printer. Here are the product's requirements:

- Must be lightweight.
- A new type of battery should be used that will provide up to four hours' continuous operation.
- Rechargeable using a standard 120 v wall plug.
- Small enough to fit into a suitcase or computer bag (may be broken down into parts which can be quickly assembled with ease).
- Target market is business travelers and students.
- Bluetooth compatible with smart phones, tablets, or computers (PC or Mac).
- Laser printer B&W or color capable (different price points).
- Price target: under $150.

Based on what we currently know, we can create a list of potential participants in our new product's development (Table 4.1).

Table 4.1 Product Development Participants.

Research and Development	Develop lightweight materials for casing and new, revolutionary print cartridge
Engineering	• Electrical – develop circuits and power supply drawings • Mechanical – design gearing, paper feed, and other components (moving and static). Determine best way to miniaturize components and how to integrate into workable product • Design packaging
Suppliers	• Battery technology • Small components, such as gears, sprockets, washers, springs, etc. • Key technology parts; e.g., Bluetooth hardware
Marketing (covered in detail in Chapter 2)	Confirm markets, arrange distribution network, determine product price, marketing collateral, packaging, product promotion
Procurement (i.e., Supply Chain)	Procurement of key items, confirmation of delivery, contracts. This may also involve warehousing regarding the storing of raw materials coming in and storage or finished products ready to go out.
Accounting and Finance	• Accounting will confirm delivery of purchased items matches with invoice, payment of invoices. • Finance will be continuously monitoring the product's finances to confirm that it is profitable (or not) and will meet margin targets.
Legal	This may be a part of the Supply Chain group, if concerned with procurement contracts, standard terms and conditions, etc. If work involved is a joint venture (JV) or acquisition of a company, it is likely the corporate attorneys will be engaged. This is generally different from the legal team in Supply Chain.
Sales	The sales team generally have a good sense of what customers desire in products and services. Sales members can offer significant insight as to product plans and target customers (similar to target market).
Quality Group	The new product's prototypes will require testing and quality standards if it is to be optimally designed for the target market. Once input by Marketing to engineering has been completed, the Quality Department will develop new testing procedures and data collection so that a quality product can be developed. QA (Quality Assurance) consists of developing the overall program and standards. QC (Quality Control) is the actual testing performed on the product.
Production (manufacturing)	Production will be enlisted to determine best approach to efficiently produce the new product. Minimal waste or scrap will be considered, with manufacturing tolerances, tooling, materials, etc.

Table developed by David Tennant.

For new products, or developing the next generation or current products, this can involve several functional areas or departments. Again, a functional area is a specialized group, such as Accounting, Engineering, Marketing, etc.

Constraint on Product Development: A Note about Sarbanes-Oxley and Publicly Held Companies

Due to accounting and financial irregularities by several large companies in the 1990s – notably Enron, Tyco, and WorldCom – Congress passed Sarbanes-Oxley (SOX), which required stronger transparency and accuracy in financial statements at publicly held companies. That is, those companies traded on stock exchanges with shares held by the public at large. As a result, this has had an impact on how accounting and procurement work at investor-owned companies. The implications for product development may be somewhat subtle. For example, if one is purchasing testing equipment for the new product, it may be necessary to follow a restrictive purchasing process and should be accounted for in the schedule. This is not to frustrate product managers, but rather to comply with Federal SOX laws.

Based on the above cases, a number of C-level executives went to prison for lengthy jail sentences. These were based on fraud, tax evasion, destroying evidence (in this case, Enron's accounting firm shredding documents), falsifying documents, and misleading investors. Several TV documentaries, books, and a movie were produced based on the Enron case (An entrepreneur figured out how to profit from scandal and misfortune ....).

Figure 4.2, Procurement Process, shows a typical procedure for how procurement is managed with publicly held companies.

Previously, one could write a purchase order (P.O.), sign off on the approval, check-in the supplier's materials at the warehouse (approving receipt), and then approve the supplier's invoice for payment – all by the same person. Since SOX became law, one will now find three different people signing off during the ordering, receipt, and invoice

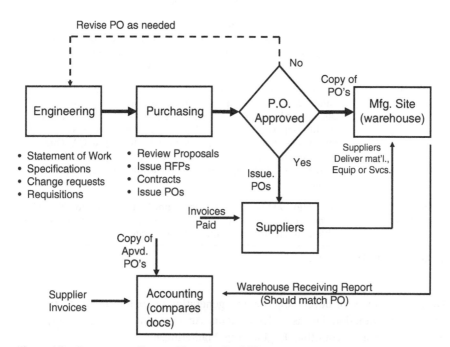

Figure 4.2 Procurement Process. Figure by David Tennant

approval process. Note that privately held companies (i.e., with no shareholders and not listed on any stock exchange) are exempt from SOX. This is because the owners would be committing fraud against themselves, not external shareholders.

Essentials of Teamwork and Communications across Functional Lines

Many companies, whether large or small, direct products or projects using a matrixed-based organization. The larger the company, the more dependent they are on the matrix. Figure 4.3 shows a typical functional organization.

During product development, the product team will "borrow" people for their team from various functional areas. This is very typical for matrixed companies. Referring to our hypothetical case (the new portable printer) we might find a product team organized as shown in Figure 4.4, Matrixed Product Team. What kind of potential conflict does this provide?

For this product team, key skilled team members were brought in from three of the firm's functional areas: Engineering, Supply Chain, and Production.

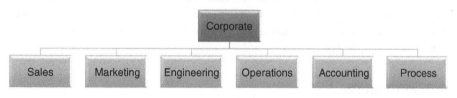

Figure 4.3 Functional Corporate Organization. Figure by David Tennant

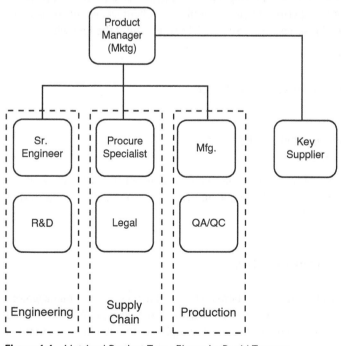

Figure 4.4 Matrixed Product Team. Figure by David Tennant

The project/product leader is from Marketing (Mktg.), another functional area. Let's consider potential issues related to this structure.

Problems with the Matrixed Team Structure

It should be noted that the team members do not report, functionally, to the Marketing Product Manager; they still report to their functional managers in Engineering, Supply Chain, and Production. If any one of the functional managers wants to pull their employee from this team, the Product Manager does not have much influence.

Similarly, if one of the team members is not performing well, the Product Manager does not have much influence other than to go to that person's functional manager.

It is with these types of issues that the supporting executive plays a crucial role. Since the Product Manager may not have the influence necessary to solve organizational issues, this is where the supporting executive can quickly resolve this type of conflict.

The matrix team structure is very common in putting project teams together. Beyond the matrix, teamwork and leadership by the product manager play a key role in team success. Some matrixed organizations have learned very well how to coordinate shared resources and motivate their product teams. But, many times the PM will find matrixed organizations are overly siloed, in which case relationships across these functional areas is a requirement. The PM will need to use these relationships to solve problems, resolve conflict, communicate status, and coordinate all the moving parts. A very tall order and not for the faint of heart.

Teamwork Essentials

Teamwork involves more than just putting a group of people in a room to implement a series of objectives. A high performing team requires excellent leadership, open communications, and shared responsibility. Successful teams exhibit the following:

- Strong sense of identity
- Shared accountability
- Trust between team members and team leader
- Enjoyment of the work
- Expressing ideas freely
- Minimal intervention by senior management
- Staying focused on end results

The team leader, or product manager, is ultimately responsible for the success of the product, and consequently must hold people on the team accountable. A large part of the PM's role is communication.

Project/Product Communication

One of the most important components of product development or project management is communication. How do we do this? What kinds of communication? What tools can I use? There are a number of techniques that can be employed. But first, it is important to point out that a high visibility project or new product will usually have an experienced leader who has a successful track record. Less senior PMs may be assigned to assist as they are being groomed to advance into product development. However, the following is a discussion of communications relative to new product development or projects.

Individual Exercise (five minutes)

List five skills that are needed for effective communication:

1.
2.
3.
4.
5.

(Answers further below)

The Product/Project Kickoff Meeting

If there is one meeting that is critical and you should prepare for, it is the kickoff meeting. Why is this important? The kickoff meeting is to accomplish the following:

- Rapport among team members
- Project/product objectives are established and confirmed
- Assign team and individual roles and responsibilities
- Begin preliminary planning (review business case, proposed scope, action items, approved budget, or schedule, etc.)
- Achieve team buy-in and team building
- Address concerns, risks, or issues the team may have
- It may be useful to have executive sponsor address team at end of session
- The kickoff meeting will set expectations

There may be additional objectives with the kickoff meeting; however, as the product leader, you only get one chance to correctly lead this important meeting. If you stumble, or are not prepared, this will set a bad tone for the meeting, and it will take you a long time to recover. In other words, you will have lost the initiative and the opportunity to make a good impression. You will have set low expectations. So, always be prepared for meetings, but especially for the kickoff meeting, which may last a full day or two, depending on the project size and complexity.

Skills needed for listening (see individual exercise above) include:

• Listening	• Focusing on speaker
• Questioning	• Informing
• Summarizing	• Persuading
• Negotiating	• Comprehending

One method the author has used to ensure everyone is clear on their role is to write on a white board (or flip chart) what each person believes is their role and responsibility. Each person in the room will describe what they will be doing on the engagement. Invariably, it becomes clear that some conflict will occur: two people may be doing (or

Table 4.2 Roles and Responsibilities: Sample Assignments.

Name/Role	Phone Ext.	Responsibility and Role
John <u>Nordin:</u> lead IT member	X-3322	Primary technical lead and decision maker for all technical aspects of project: • Responsible for analyzing suppliers and technical solutions. • In charge of design team. • Develops standards and testing criteria. • Ensures technical solutions meet requirements.
Marie Santello: Procurement	X-4117	Liaison between project team and procurement department: • Works with project team to procure needed hardware. • Ensures that company purchasing policies are followed. • Expedites at-risk material deliveries. • Develops procurement schedule. • Assist with RFP and bid-award process.
Jose Perez: project manager	X-5434	Overall responsibility and authority for project implementation: • Directs and manages the project team. • Serves as link between manufacturing, the project team, and executive management. • Develops and manages scope, budget, schedule.

Table by David Tennant

think they are doing) the same tasks, others may try to shift work to others, etc. It is the job of the product leader to facilitate this discussion and ensure everyone is clear on their roles at the end of the day.

Table 4.2, Roles and Responsibilities, illustrates how these can be captured for the team. This sample shows only three people; however, each member of the team should have an assignment. It is expected that a full team R&R chart would contain 10 to 12 people.

Budgets, Schedules, and Miscellaneous Small Tasks

It is worth noting that budgets and schedules in a business case do not mean much. It is, at that point, simply an educated guess. It is likely that supplier quotes have not been obtained and the scope (breadth) of the work has not been developed in any detail. The REAL scope, budget, and schedule will be determined by the product team, with input from a variety of sources, including external suppliers.

Budgets and cash flows will be discussed further in this chapter and schedules are discussed further in this book. However, there will be many day-to-day activities that are important – we don't want them to fall through the cracks – but they may not be significant enough to place into a product schedule.

One mechanism to track all the ongoing small items is to use an Excel spreadsheet to create an Action Item Log. This is an extremely useful tool whether used by a small three-person firm or a large Fortune 100 company. Table 4.3 Action Item Log is a

Table 4.3 Action Item Log Sample.

Item	Description	Assigned To:	Date	Status	Expected Resolution Date	Notes	Priority
1	Approved supplier list needs to be revised. New suppliers need to be added. Engineering requests this ASAP	MS/RB	29/11/20	Open	03/01/21	One supplier has been slow to respond. Need to agressively follow up, need equipment list with updated delivery dates.	H
2	ABC to provide a complete list of supporting equipment and identify those that include a factory acceptance test. AC pointed out that the factory acceptance test is part of the new facility commissioning process.	LR	13/12/20	Open	07/01/21		M
3	Review of casing materials and screening of suppliers needs to occur in the next 30 to 45 days.	KB	21/01/21	Open	15/02/21	Meet with POC to explore this. End users do not require stainless steel material. Alternate pricing required from Vendors for variety of appropriate materials for harsh environment, including rugged plastics.	H
4	Sub supplier owes Form 1 of QC program.	BB	15/03/21	Closed	01/04/21	Follow up needed.	L
5	F.P. working on sketch for conduit routing. Beta client requests 7'-10" headroom; but current best est. is 7'-3."	FP/RB	01/04/21	In progress	10/04/21	Follow up with Ken. Sketches received. Meet with Engr to discuss.	M
6	Obtain all supplier invoices showing sales taxes paid	RB/MS	30/03/21	Hold	Hold	Waiting for direction from Acctg. They will check with State for documentation requirements.	M
7	Catwalk issues at test site .... adequate strength for installation equipment?	AC	25/03/21	In progress	10/04/21	See memo 1/30/21 ABC to provide updated schedule for catwalks. Alternate methods may become necessary. Earliest start is June 15, 2021.	H
8	All sketches to be confirmed and turned over to Engr.	RB/MS	05/03/21	In progress	07/03/21	Most sketches received. See R.T. on this item	H
9	Prototype test equipment needed at client beta site.	RB/NJ	01/07/21	In progress	10/07/21	QA/QC lab to coordinate with field techs.	H

Table developed by David Tennant

spreadsheet example. This tool can be discussed during status meetings: What's been completed? What's still open? Which new items need to be added? Those activities that have been completed can be moved to a second spreadsheet labeled "Closed."

At the end of the project, you will have a complete listing of all "action items," when they were completed, by whom, status, and any notes. This can also be used to document key decisions.

Note in the above table that we can also assign a Priority: H, M, L (High, Medium, Low).

Status Reporting

Most projects have a weekly status meeting. This is the opportunity to discuss issues and walk through the Action Item list. This should be a quick overview. A monthly status report should be issued that covers:

- Events over the past 30 days
- Current challenges
- Tasks to be tackled over the next 30 days
- Brief discussion of budget and schedule

Note, if a project is only a few months in duration, it may be useful to have more frequent meetings and status reports. The author has seen status reports up to 50 pages in length. These are for large dollar projects or programs (usually $50 million and up). However, who will read a 50-page report? These specific reports had full budgets and schedules embedded, along with project photos, and other documentation – all of which is unnecessary.

The above bullet points should be covered in around two pages maximum. It is acknowledged that some clients will want more detail. However, keep the report limited to two pages and provide further documentation as an attachment or appendix. NO ONE will read a 50-page status report.

Presentations

It has been stated that most people would rather have a root canal treatment than give a presentation. However, it will be important for you as a product manager to be comfortable giving presentations. This is true regardless of the size your organization: Fortune 100 company, single entrepreneur, or small business. You may be asked to give a presentation for the following:

- Status – The senior management of your company want a quarterly presentation of the project's status. This may even include your firm's Board of Directors.
- Investor Groups, Incubators, or Banks – Should you need funding or seek an incubator for assistance, you will need to clearly pitch what your company, product, or idea is about. Note, investor groups and incubators are discussed at length in Chapter 6.
- Civic or Community Groups – This might be presentations on your new product to Rotary Clubs, Chambers of Commerce, business associations, etc.

- Mass Media – Should your new product or idea catch the attention of the media, you may be asked to do an interview. Think of Elon Musk, Richard Branson, and other high visibility people.
- Conferences – To promote your new product, presentations at conferences and trade shows may be appropriate.

Do the above forums make you nervous? Would you be capable of putting together an impactful presentation and deliver smoothly?

Everyone has varying degrees of comfort or experience in giving presentations. Unfortunately, the only way to get better at presentations, is to do them. Regardless of your current skill level, it is highly suggested that, in leadership positions, you will be expected to make presentations. If this is one of your weak points, there are many companies that offer courses in public speaking and many chapters of Toastmasters™ are available nationwide. Toastmaster's International™ website offers the opportunity to look up local chapters.

https://www.toastmasters.org

This book does not cover the art of putting together presentations or how to utilize Microsoft PowerPoint™. But there are many tools, excellent books, and YouTube videos on putting together presentations. As a product manager, or in _any_ role, giving effective presentations should be one of the tools in your toolbox.

Leadership in Product Development

Leadership is a vague and broad topic. What is leadership really? Can anyone develop their leadership skills? How does this relate to product development or project management?

First of all, leadership skills are necessary to encourage, direct, and motivate teams. It is needed to interact with and persuade people of influence (think senior management) and to negotiate with suppliers or other functional areas. There are many definitions of leadership; a few are listed below:

> A Leader takes people where they would never go on their own.[1]
>
> –Hans Finzel

> Leadership is influence, period.[2]
>
> – John Maxwell

> Key elements of leadership include vision, character, commitment, leading change.[3]
>
> – Dave Ulrich

Pick up any book on leadership and you will find a leader defined as someone:

- Having high personal integrity (Usually #1 on the list)
- Visionary
- Focused on Results

- Having strong interpersonal skills
- A great communicator
- Able to motivate others
- A change agent
- Follows through on commitments
- The ability to influence others
- Getting things done through other people
- Capable and confident.

New product development means working with and leading others in a team. Can we improve our leaderships skills? The answer is yes. However, it is noted that some people are born with natural charisma; others may simply have an engaging personality style, and not everyone can develop this. However, each of us can improve our leadership skills. Some of this is by learned behavior, observation, and taking specific steps for improvement. These could include taking courses in negotiating, conflict resolution, public speaking, decision making, and other topics.

Books on leadership, as demonstrated above, have several recurring themes. However, two of these that consistently appear are integrity and trust. If your team feels they cannot trust you, they will not follow you. Further, senior management will not give you authority or responsibility if they feel you lack integrity. It is a necessary ingredient for leadership.

Diving a little deeper on a few of the above bullets:

Visionary. This not the easiest perspective to visualize what the end result or product will be. And, how to communicate the "vision" to the team. Sometimes, C-level leaders have challenges communicating their vision or end result to the team or employees. Steven Covey, in his book "The Seven Habits of Highly Effective People," has a chapter entitled: "Begin with the End in Mind." He goes on:

> Begin with the end in mind is based on the principle that all things are created twice. There's a mental or first creation, and a physical or second creation to all things.[4]

Change Agent. A leader wants to implement change, they are never happy with the status quo. Is your new product or idea a disruptive technology? Will it revolutionize the way people look at your product or its application? Whether promoting new products or ways of doing things, or changing a company's culture, leaders are always looking for a better way to do things, to implement change.

Follow Through on Commitments. Have you ever assigned a task to someone, with a deadline, and that person missed it? Or they gave you a less than satisfactory deliverable? What was your view of this person? Would you trust them again with an important assignment? If you make a commitment to your boss, your team, or others, be sure to fulfill your assignment. If you think you will be late or need assistance, communicate this; most people are reasonable and will offer help.

Getting Things Done Through Other People. For the most part, you cannot do everything. That's why there's a team. There will be times when you need to delegate

Table 4.4 Approaches to Work.

	Leadership	**Management**
Planning	Strategies, vision, providing resources, funding, support, relationships	Develop budgets and schedules, staffing, confirm objectives, direct team
Execution	Motivation, resolve high-level conflict, funding, people relationships, distribute information, obtain feedback	Resolve problems, control efforts, report status, assign tasks

Table by David Tennant

key tasks or activities to others. All good leaders learn the art of delegation. The key is to make sure the person you delegate to has the necessary skill set and commitment.

Capable and Confident. Successful leaders exude an aura of confidence. Leaders must have confidence in their abilities and the purpose of their vision. In developing and promoting your new product, do you believe in it? Do you feel confident in your abilities to bring it to fruition?

There is also a difference in the types of work performed by a leader vs. a manager (Table 4.4).

In his book, "The Five Dysfunctions of Teams," Patrick Lencioni[5] identifies the following:

1. Absence of Trust
2. Fear of Conflict
3. Lack of Commitment
4. Avoidance of Accountability
5. Inattention to Results.

Most of the above are self-explanatory, but further discussion is needed for a few.

Fear of Conflict. People who avoid conflict do not make good leaders. In the early stages of team formation, there will always be conflict. This means constructive conflict, such as disagreements on approach, differences on setting priorities, budgets, etc. A good leader allows these discussions to continue (within a reasonable time frame). However, when conflict starts to become acrimonious, such as name calling, hard feelings, disruptive behavior, etc. a leader must step in *quickly* to diffuse the situation.

Avoidance of Accountability. The people on your team should he held accountable. Are deadlines being met? Did that contract get issued to the supplier on time? If people are not performing, your team is looking for you to fix the problem. Leaders are also held accountable: did the product get released on time? Are there production problems?

How Do Leaders Go Wrong?

Is ego a bad thing? To reach the levels of influence and to push your ideas along, a certain amount of ego is necessary. It is when ego gets out of control that things go wrong. How many times has someone been put in charge for the first time, and it goes to their head? Below is a listing of how leaders go off track:

- Abusive and arrogant
- Dictatorship in decision making
- Egocentric manner
- Poor interpersonal skills
- Absence of praise or affirmation
- Lack of delegation
- Poor communicator
- Oblivious to the corporate culture
- Poor listener.

A few of these bear further explanation:

Dictatorship in Decision Making. This means not allowing other opinions, or dissent, in making key decisions. This is the opposite of open and honest teamwork and defines a leader with a larger than needed ego. These leaders are convinced of their invincibility and that they alone hold all the keys. It is always a good idea to solicit other opinions for important decisions. Four or five heads working together will always be better than one.

Poor Interpersonal Skills. As a leader, you should understand empathy, self-awareness, and emotional intelligence. Yet, the author has worked with several high-level people who lacked some or all of these qualities. Having empathy means you can sense other people's emotions and what they may be feeling. Self-awareness is knowing your own abilities, limitations, and emotions and how that may influence your behavior. And finally, emotional intelligence allows one to recognize, understand, and respond to the emotions of others. These are key ingredients that are needed to be a truly successful leader.

Absence of Praise or Affirmation. People generally like to be acknowledged for having done an excellent job or task. Thanks for a job well done gives people confidence, encourages them to do their best, and build morale for the team. The lack of affirmation can obviously do the opposite by discouraging people. Always provide praise for those doing an excellent job; or where appropriate, recognize the team.

Oblivious to the Corporate Culture. Corporate culture is one of those items that is hard to define. Basically, it is how people within the corporation behave and interact with each other, their management, and customers. It is an unwritten set of expected behaviors. For example, high tech firms, such as Google™ or Apple™ have a more relaxed culture than, for example, utilities or banks. Dress codes are different and employees at high tech firms may find pool tables and beer in the break room. These are different from traditional work environments and leaders who are brought into a firm need to be aware of the corporate culture. If the culture (behaviors) needs changing, this will not happen overnight.

In summary, strong teamwork and leadership must work together to be successful. The implications are that new product development (a project) will be more efficient, open communications will allow problems to emerge more quickly (and solved earlier), and the efforts of a high performing team should translate into a better product and profitability.

Tabletop Discussion

1. What can I do personally to improve my own leadership skills?
2. What are some of the traits I have observed in leaders that I admire?
3. What will you do to develop people around you (or who work for you)?
4. What are some actions you can take as a leader to build morale?
5. How will you hold people accountable?

One technique is to observe leaders and mimic their behavior. For example, there are many videos of highly influential people on the internet giving speeches and presentations. This includes business leaders, military and political figures, religious and community leaders. Watch their body movements (hands, facial expressions, posture) and observe their speaking style. Do they appear nervous? Anxious? Passionate? Excited? How are they dressed? One can learn a lot by observation – to the point where you can use their techniques to your advantage.

Similarly, observe leaders in your company or community.

The Roles of Accounting and Finance

The accounting department at any company has three primary responsibilities:

- Accounts Receivable – obtaining payment from clients or customers (cash in)
- Accounts Payable – providing payment to suppliers, employees, taxes, etc. (cash out)
- Developing reports, such as an annual report, internal budgets, and cost-tracking.

Note that accounting is not responsible for tracking in detail individual project or product development costs. This usually falls to the product manager and creates different issues for the product development team.

Typically, a PM will receive accounting reports that indicate the dollar amounts expended and the amount of funding left in the product's account. This is different from tracking a budget and schedule in detail, which is what a product/project manager requires. To a large degree, the PM must develop a schedule and supporting budget spreadsheets to accurately track a product's development.

A Sample Product Budget

Table 4.5 is a representation of a product budget. The development effort is scheduled for 8 months and will cost just under $8 million. Note that this is also a cash flow diagram, shown in graphic form in Figure 4.5. This cash flow illustrates the dollars required on a monthly basis. It will be possible for the PM to compare monthly, how much funds are expended compared against the cash flow plan.

Figure 4.5 illustrates the cashflow graph for this project.

Table 4.5 Sample Budget.

Month	TTLs	1	2	3	4	5	6	7	8
New Widget Product Budget									
Capital investment – New Mfg Line									
Numeric control machines (4)	$4,000,000			$400,000					$3,600,000
Building modifications									
Electrical	$696,500	$20,000	$34,500	$50,000	$147,000	$175,000	$200,000	$50,000	$20,000
Mechanical	$227,000	$-	$-	$35,000	$55,000	$60,000	$60,000	$12,000	$5,000
Packaging	$26,500				$25,000				$1,500
Purchase testing equipment	$200,000	$-	$-	$75,000	$125,000	$-	$-	$-	$-
Engineering – Product Dev	$600,000	$50,000	$50,000	$100,000	$100,000	$75,000	$75,000	$75,000	$75,000
Elect – product circuit design	$230,000	$-	$-	$35,000	$35,000	$40,000	$40,000	$40,000	$40,000
R&D	$480,000								
Project Team	$480,000	$60,000	$60,000	$60,000	$60,000	$60,000	$60,000	$60,000	$60,000
Supplier raw materials	$142,000	$20,000	$20,000	$20,000	$20,000	$30,000	$20,000	$12,000	$-
Prototype development	$208,000	$10,000	$24,000	$24,000	$15,000	$15,000	$40,000	$40,000	$40,000
Testing and data collection	$60,000	$-	$-	$10,000	$10,000	$10,000	$10,000	$10,000	$10,000
Technicians	$80,000	$10,000	$10,000	$10,000	$10,000	$10,000	$10,000	$10,000	$10,000
Overhead	$371,000	$24,000	$27,000	$52,000	$72,000	$58,000	$63,000	$40,000	$35,000
Totals	**$7,801,000**	**$254,000**	**$285,500**	**$931,000**	**$734,000**	**$593,000**	**$638,000**	**$409,000**	**$3,956,500**

Table developed by David Tennant

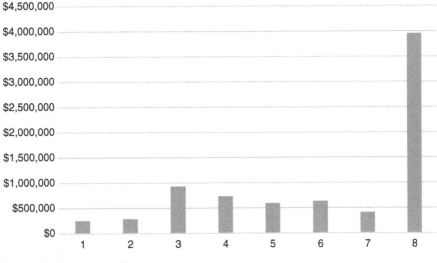

Figure 4.5 Product Cash Flow.

Pro Forma Modeling

Most companies will develop a proposed cashflow model or proforma that provides a visual means to determine margins and payback over a specified timeframe as shown in Table 4.6 which represents a new Software deployment to customers.

Note that this pro forma represents one year of costs and revenues and they are estimates. In this instance we make the following assumptions:

- The company did a feasibility study
- Development and testing successfully occurred in January and February
- Marketing did adequate research and SWOT review
- Sales teams are ready to sell to customers
- Pricing has been determined through market research
- Distribution channels are in place
- An advertising campaign is ready to launch.

Note that the number of customers who are interested in and ultimately buy the product are estimates. The costs, revenues, and profits (net income) are based on these estimates. Based on projections in Table 4.6, breakeven occurs in month three (going from a loss to a profit).

1. What would the net income look like if the revenue (i.e., customer count) is lower than expected? What effect would this have on costs?
2. What if the product is wildly successful? How would this impact the company if the number of customers was significantly higher?

If the customer count is lower, it is likely that some cost would be lower (travel, legal, cost of sales, etc.). However, this may infer that more money needs to be expended in marketing and sales efforts to boost revenue. It would also likely shift the breakeven point further out. Note also that this product has very high margins; $53 million

Table 4.6 Sample Proforma Developed by David Tennant.

Product Zed Software and Services Proforma

Item	Jan	Feb	Mar	Apr	May	Jun	Jul	Aug	Sep	Oct	Nov	Dec	TTLs
No. of New Customers	0	0	10	10	20	25	40	60	70	100	100	100	
Gross Revenue													
Basic Software package = $55,000			$550,000	$550,000	$1,100,000	$1,375,000	$2,200,000	$3,300,000	$3,850,000	$5,500,000	$5,500,000	$5,500,000	$29,425,000
Deployment (consulting) = $20,000			$200,000	$200,000	$400,000	$500,000	$800,000	$1,200,000	$1,400,000	$2,000,000	$2,000,000	$2,000,000	$10,700,000
Add-on Optional Modules = $15,000	$-	$-	$150,000	$150,000	$300,000	$375,000	$600,000	$900,000	$1,050,000	$1,500,000	$1,500,000	$1,500,000	$8,025,000
Customer Training (per cost) = $35,000			$350,000	$350,000	$700,000	$875,000	$1,400,000	$2,100,000	$2,450,000	$3,500,000	$3,500,000	$3,500,000	$18,725,000
Gross Revenue	$-	$-	$1,250,000	$1,250,000	$2,500,000	$3,125,000	$5,000,000	$7,500,000	$8,750,000	$12,500,000	$12,500,000	$12,500,000	$66,875,000
Expenses A&G													
Special test equipment	$375,000	$-	$-										$375,000
Software Deployment – Subset	$50,000	$68,000	$250,000	$400,000	$400,000	$400,000	$450,000	$500,000	$550,000	$650,000	$650,000	$650,000	$5,018,000
CIA/QC of new Software Testing	$75,000	$125,000	$-										$200,000
Internal Sales and Technician Training	$20,000	$50,000	$50,000	$50,000	$50,000								$220,000
Customer Beta Test and Deployment	$25,000	$25,000	$-	$-	$-	$-	$-	$-					$50,000
Travel Expenses	$-	$-	$50,000	$100,000	$100,000	$150,000	$200,000	$250,000	$300,000	$450,000	$450,000	$450,000	$2,500,000
Marketing and PR	$250,000	$250,000	$200,000	$100,000	$100,000	$100,000	$100,000	$100,000	$100,000	$100,000	$100,000	$100,000	$1,600,000
External Consultants	$80,000	$100,000	$125,000	$50,000	$10,500	$-	$-	$-	$-	$-	$-	$-	$365,500
Legal – Contracts, guidance	$3,500	$3,000	$25,000	$25,000	$25,000	$25,000	$25,000	$25,000	$25,000	$25,000	$25,000	$25,000	$256,500
Trade shows, conferences, webinars	$100,000	$100,000	$100,000	$100,000	$100,000	$100,000	$100,000	$100,000	$100,000	$100,000			$1,00,000
Cost of sales, delivery, etc.	$125,000	$125,000	$125,000	$125,000	$125,000	$200,000	$200,000	$250,000	$250,000	$250,000	$250,000	$250,000	$2,275,000
Total Expenses	$1,103,500	$846,000	$925,000	$950,000	$910,500	$975,000	$1,075,000	$1,225,000	$1,325,000	$1,575,000	$1,475,000	$1,475,000	$13,860,000
Net Income	$(1,103,500)	$(846,000)	$325,000	$300,000	$1,589,500	$2,150,000	$3,925,000	$6,275,000	$7,425,000	$10,925,000	$11,025,000	$11,025,000	$53,015,000

estimated for the first year. It would take a significant downturn in sales before this product would be a loser.

If the sales numbers were much higher, this could also have a negative impact. If sales were double the projections, would there be enough staff to support twice the deployments? Would the company need to quickly hire more technicians and software engineers to support this effort?

It is likely the company would perform sensitivity analysis to determine costs, profits, and support staff needed based on both less and more customers. It should be apparent that both marketing and sales are necessary for product success.

Decision Points and Net Present Value (NPV)

It is important to remember that companies are in business to make a profit. If profitability is missing, the company will ultimately go out of business: no employees or payroll, no taxes paid to local governments, and less economic activity in the business' community.

Throughout the development of a new product, there will be strategic reviews or "gates" through which the project must pass to continue funding, as described earlier. One yardstick that measure a product's viability is net present value (NPV). In NPV analysis, we take all the future cash flows (in and out) and discount them to today's cost. This allows one to compare several projects (or potential products) and select the most profitable scenario.

The NPV relies on a discount rate, required company return and inflation that is based on the cost of the capital required to make the investment. Note that investment in this context could be a project, new product, company acquisition, or joint venture with another firm. Also, is it important to realize that any NPV analysis makes assumptions about the future that may be unreliable.

A few points to keep in mind:

All companies have a required rate of return. Does your company require a 20% return on investments (ROI)? 25%? 30%? More? Some companies may want to take inflation into account. 2%? 3%? 5%? And, if your company is borrowing the funds, the cost of capital (interest rate) would require inclusion.

Suppose a company wants to invest in one of three products/projects. Each project is estimated at $2 million, but your firm only has enough money to invest in one of them. How would we choose which product to invest in? NPV analysis can answer this from a purely financial perspective.

Let's look at a potential investment and first construct a cash flow table (Table 4.7) based on a six-year timeframe:

The cashflows out are negative as this is money your company is spending on the development of a new product.

In year three, we are starting to see income (i.e., revenue) from our new product, which is represented by a positive number. The difference between the outflow and inflow for each year is the net income, which is negative for the first three years. However, is this product meeting our company's required Return on Investment (ROI)? Assume that our company requires a 15% return on investment, and we also want to consider inflation over six years. How would we determine this? The answer is a NPV evaluation.

Table 4.7 Simple Cash Flow.

New Product Cashflows

Year	Outflow	Inflow	Net
1	$ (2,500,000)	$0	$ (2,500,000)
2	$ (5,000,000)	$0	$ (5,000,000)
3	$ (1,500,000)	$ 400,000	$ (1,100,000)
4	$0	$ 3,500,000	$ 3,500,000
5	$0	$ 5,500,000	$ 5,500,000
6	$0	$ 6,000,000	$ 6,000,000
TTLs	$ (9,000,000)	$ 15,400,000	$ 6,400,000

Table Developed by David Tennant

Table 4.8 NPV Calculation.

Year	Outflow	Inflow	Net	Discount Factor*	NPV
1	$ (2,500,000)	$0	$ (2,500,000)	1.0000	$ (2,500,000)
2	$ (5,000,000)	$0	$ (5,000,000)	0.8547	$ (4,273,504)
3	$ (1,500,000)	$ 400,000	$ (1,100,000)	0.7305	$ (803,565)
4	$0	$ 3,500,000	$ 3,500,000	0.6244	$ 2,185,297
5	$0	$ 5,500,000	$ 5,500,000	0.5337	$ 2,935,075
6	$0	$ 6,000,000	$ 6,000,000	0.4561	$ 2,736,667
TTLs	$ (9,000,000)	$ 15,400,000	$ 6,400,000		$ 279,970

* discount factor = $1/(1 + k + p)$ exp t
k = ROI = 15%; p = inflation = 2%
Table developed by David Tennant

Table 4.8 takes the information in Table 4.7 and performs further analysis to determine the Net Present Value of this product's development. That is, it is taking the estimated future cash flows, and bringing them back to today's value.

Notes about Table 4.8

Our company's required rate of return (same as ROI, in this case) is 15%. We can also account for inflation for an average of 2% over the next two years. The "t" equals the year and is an exponent.

In the formula, t = the year. This is needed to calculate the discount factor for each year. In this table:

Year one, t = 0
Year two, t = 1

Year three, t = 2
Year four, t = 3
Year five, t = 4
Year six, t = 5

Remember from algebra that any power raised to zero = 1. So here are the first three years of calculated discount formulas:

$1/(1 + .15 + .02)^t = 1/(1.17)^t$
Year 1: $1/(1.17)^0 = 1/1 = 1$
Year 2: $1/(1.17)^1 = 1/1.17 = 0.8547$
Year 3: $1/(1.17)^2 = 1/1.3689 = 0.7305$

Year four, five and six follow the same logic which is how we get the numbers in Figure 4.6. Note that the total NPV = $279,970. What does this number mean? Consider the following:

If the company was getting exactly 17% (15% for required return plus 2% for inflation), the total NPV would equal zero (NPV = 0). Anything greater than zero means that the company is making more than 15%. Since $279,970 is larger than zero, this product will make over 15%.

If the total NPV was a negative number (i.e., NPV < 0), this would mean the company is making less that it's required 15% return.

Some notes on Net Present Value:

The NPV is based on assumptions, which could be wrong. For example, the inflows might be less, (or more) depending on sales (cash inflow). Inflation might go higher, which could also affect the NPV.

What do we do if our total NPV is negative? A number of things: we can stretch out the time from six years to eight years or longer. That might help, and we can also look at cutting costs. But note that "Figures don't lie, but liars do figure." The author has seen instances where an analyst will play with the numbers until the "right" NPV is reached (zero or higher). Be aware of what assumptions are used in developing the NPV. Usually, companies will do a sensitivity analysis to determine what the NPV will be over several scenarios.

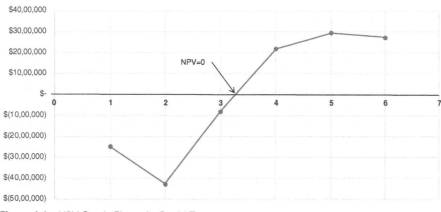

Figure 4.6 NPV Graph. Figure by David Tennant

Table 4.9 New Product Selection.

Project No.	Investment Required	Dev. Time Frame (Yrs) I	Risk Rank 1 = low 10 = high	NPV	Expected Payback (yrs)
1	$5,000,000	1.5	2	325	4
2	$3,250,100	2	3	401	3
3	$15,000,500	3	8	130	4
4	$12,725,000	5	9	390	5.5
5	$7,565,000	4	5	121	3

Table Developed by David Tennant

This NPV example was presented so the reader can understand how NPV is calculated and what it means. NPV can be easily calculated using Excel functions, eliminating the need to construct tables. Some hand-held business calculators will also do this function.

It is also important to note that NPV analysis is extremely useful in comparing multiple projects. Consider Table 4.9 New Product Selection. Our senior team has performed an analysis of several potential new products. Suppose your company has $12 million to invest in new product development projects. Based on this table which products should your company invest?

There are a number of factors to consider in addition to NPV in this example.

With only $12 million available, this would eliminate projects three and four. Project 4 would be a stretch, but it is likely that an additional $725,000 could be found, or if necessary borrowing the funds could be arranged. However, in this case, the company has done a risk review to assess their chances of success. The two most expensive products on this list are high risks and have the longest payback times.

Therefore, for this company it would make sense to invest their dollars in projects one and two. Both products are low risk, have reasonably quick payback times and the NPVs are healthy. Both projects would come in under the $12 million threshold and would be profitable based on net present values. Project five, while under the investment threshold, has the lowest NPV and therefore, is eliminated.

The Bigger Picture

The previous discussions about budgets, cash flows, and NPV are extremely important in new product development. First and foremost, companies rely on NPV to determine if a product is worth pursuing or continuing. Once the project is put into motion it is important that the NPV analysis remain positive at all gate or review points. One way to keep the NPV positive is to closely monitor and control the budget. If your project starts to go over budget (additional cash out), this will impact the project's NPV. Several points to keep in mind:

- It is important that your new product's anticipated development costs and schedule are realistic. This will be determined by the scope assigned to your project and how deftly the PM controls the budget.

- It is important that the project team develop an accurate cash flow table.
- If the project goes significantly over budget and NPV becomes negative, there is a good chance upper management will cancel the project.
- NPV analysis considers the time value of money.

It cannot be stated more clearly that the Product Manager must manage the budget and minimize product changes (i.e., be profitable), which add to the overall costs. If development costs are out of control, the new product will never succeed. Note that Agile Project Management, primarily an application for software and systems products, thrives on continuous change. Agile is discussed further in Chapter 5.

What are some of the circumstances under which products in development get canceled?

- Change in leadership at the company. A change in top leadership means that new priorities will be set, some projects (or products) canceled, and different new ones put in motion.
- A competitor beats your company to market. This should illustrate that time and schedules are important. Time is also money.
- The product/project goes significantly over budget with NPV showing a negative return.
- Technical obsolescence. Technology changes so quickly in today's world that new products may be technically or functionally obsolete before development is completed.
- The regulatory environment can change. For example, environmental concerns are pushing public policy away from internal combustion engines (found in all cars and trucks) and towards clean energy fuels such as electric vehicles, natural gas or, potentially hydrogen. Some products live and die by regulations.

Do products/projects really get canceled? The answer is yes, more than you think.

Driving Product Development

Who is best suited to lead the product development effort? Which functional area should be in charge: Marketing? R&D? Engineering?

One of the three functional areas above will typically drive the development effort. Most commonly, marketing is the probable candidate. But, besides the functional area and market knowledge, what other aspects are important? The author has seen numerous product development efforts and can identify several personal characteristics for success.

1. Experience in managing projects. The failure in products or projects is many times due to a lack of detailed planning and execution methodology and inexperienced project leaders. This is a key success factor in any project. Without this expertise, it is likely that budget and schedule targets will be missed, communications will be poor, and "firefighting" will be a common practice.
2. Trust and Integrity. No one likes to work for an overbearing dictator or micro manager. Also, those with huge egos may intimidate people. These styles stifle creativity, suppress open discussion and ideas, and can drive the best people away.

   The most effective product or project leaders are those who provide guidance, encourage trust and open communications, provide the tools or resources necessary for success, and make key decisions by consensus (with input from the team).

Additionally, they are knowledgeable in the core principles of project management (discussed in another chapter).

3. Strong communication skills. Poor communication skills can create misunderstandings and conflict. Effective leaders are comfortable in writing memos and reports, and in presenting. Instructions should be clear, concise, and easily understood.

4. Is comfortable in building relationships. To get things done and coordinate with other departments or functional areas, the PM must be adept at approaching people to ask for assistance (and this is a two-way street, offer assistance when needed), regardless of the level (junior manager up to the C-suite). Building relationships is important and a skill all leaders possess.

Tabletop Discussion

1. Think about the last time you enjoyed working on a team. What made it enjoyable?
2. Think about the team leader on that assignment. What attributes did he or she possess that you found attractive?
3. Does the team leader need to be a technical expert?
4. What strengths do you possess that would be good for product development and team leadership?

Working in Silos and with Stakeholders

As mentioned earlier in this chapter, companies are organized by functional areas. And, depending on the corporate culture, communications across those areas can be difficult. Further, some functional managers may be overly protective of their "turf." In any project, and especially new product development, it is necessary that communications and operations are not impeded.

Discussion Case 4.1 – Organizational Conflict: A True Case Study

This Fortune 100 company had anywhere from 200 to 400 technical projects ongoing at any time. Also included were around five or six new products in development. The company, headquartered in the Southeast United States, has over 27,000 employees and has annual revenue of $20 billion. Some of the projects were small (duration: one to three months, under $250,000) and others, very large (duration: over four months to 24 months and millions of dollars). The company had a large engineering organization. Two of the managers within the engineering department disliked each other and each did their best to antagonize the other.

In one instance, the manager of electrical engineering accused the manager of instrumentation and controls (I&C) of "grabbing his work." There was some truth to this, and the I&C manager was deliberately probing as he knew this would irritate the other manager. In some companies, I&C and electrical engineering are closely related and may even be within the same department discipline or design group. In this company, they are separate, distinct groups. They were both supporting a new product development effort and their in-fighting was taking a toll on the team as well as putting the product behind schedule.

The organization was highly siloed and coordination between groups was unpredictable. Assume you are the Product Management leader. Consider the following:

1. How would you handle this ongoing, organizational conflict?
2. What are your chances of getting the two managers to work with each other?
3. Since this is an organizational issue, how can we resolve this?
4. Is this a case of personality conflict?
5. What does your team expect of you, the product manager?

Poor communications aggravated by differing personality styles, can derail the best of intentions and new products. As the product leader, your team is looking to you for leadership in resolving conflict, providing inspiration, clear communications, providing supporting resources, and staying focused on results.

Working in silos refers to employees working strictly within their department with minimal communications or coordination outside of their department. This is clearly not the best scenario for the matrixed organization where teams are formed from various functional areas. This is an organizational issue and not one that a PM can resolve at his or her level. However, as the project/product leader, the PM is expected to reach across silo barriers and coordinate the various groups. It should be one of the PMs fundamental competencies.

Identifying Stakeholders

What is a stakeholder? The classic definition is anyone who can impact your project (or product) or be impacted by your project (or product).

Developing a product or service is a project. It involves a team working toward a common goal, a sponsor (usually providing funding), end users, distributors, and suppliers. However, it should be noted that some stakeholders are more important. For example, the C-suite (CFO, CEO, COO, etc.) may be very interested and engaged with your new product's development. Perhaps the supply chain group, is playing a minor role. However, be assured that it is necessary to identify all the stakeholders that are involved in your development effort.

What happens if we overlook an important stakeholder? When the stakeholder (perhaps a high-level Director or VP) finds out, he or she will want changes. "Why wasn't I informed of this?" "I want to see some additional features." This will mean changes to the product's design, which will impact the overall cost and schedule – most likely in a negative fashion.

What can we do to identify all stakeholders? There are several techniques:

- Determine who is funding the project and ask him or her who are the key players.
- Look at the WBS (Work Breakdown Structure) to determine which functional areas and suppliers are involved (WBS is described in a subsequent chapter).
- Go to each functional area that you know is engaged in the product and ask questions

Once the stakeholders are identified, a table showing the stakeholders can be developed. Table 4.10 is a sample stakeholder listing.

Table 4.10 Stakeholders in Project X.

Name	Title	Role	Influence
J. Edwards	Sr. VP Marketing	Sponsor	High
M. Dowd	Mgr. Engineering	Product Design	High
M. Henry	VP sales	Product Distribution	High
B. Johnson	Supplier	Brochure Development	Low
J. Perez	Supplier	Test Equipment	Low
L. Cincotti	Sr. Buyer	Procurement	Med
G. Hinson	Plant Manager	Manufacturing	High
L. McCarty	Mgr. Quality	Product testing	Med
J. Nelson	Corp. Attorney	Legal – review contracts	Med
B. Walcott	Mgr. Contracts	Supplier – distribution	High

Table developed by David Tennant

Note that those stakeholders with a high level of influence should be kept fully informed of product status through meetings, status reports, and other means. In other words, you should do a lot of "hand holding." Those with medium to low influence are probably not kept informed to the same degree but recognize that their participation in the project is still important.

A part of your role as the Product Manager is to manage the stakeholders. Note that this entails a lot of relationship building. If you do not manage the stakeholders, they will manage you and you do not want four or five stakeholders directing your efforts. It is important that relationships are formed, and key stakeholders' requirements and expectations are made known.

Note that requirements (technical, commercial, functional, etc.) are generally easy to determine. It is the expectations that are the hard part. What are expectations? It includes those intangible items that usually aren't stated or obvious. For example, perhaps the sales team is expecting a light-weight gizmo. The engineering manager envisions a robust and weighty case based on expected use. The engineering and sales teams may have different ideas about the final product but haven't specifically stated this. This could further translate to color of the product, shape, differentiation of market tastes, etc.

It is important that expectations are expressed so that these can be incorporated into the product's design features. How do we determine expectations? Relationships.

Chapter Key Points

- All projects, including product development, must support the company's strategic plan.
- The sponsoring executive provides funding, support, and resolves conflict that the team cannot resolve on its own.
- Project management principles are needed to plan, design, develop, and launch a new product or project.

- In developing a new product, there will be a variety of functional areas involved.
- Sarbanes-Oxley has had an influence on how accounting, finance, supply chain, and almost every functional area operates in a publicly held company. Ensure that extra time (and possibly budget) is allocated to your project.
- Teamwork is vital to product/project success.
- It is necessary that the product or project manager monitors and controls the budget and schedule. If the budget is overrun significantly (making the NPV negative), the new product or project may be canceled.
- NPV (net present value) is a measure of a project's profitability.
- It is important that all stakeholders are identified, and relationship-building occurs with the highly influential ones.
- The product/project leader must keep the "big picture" in mind: coordinating budgets and schedules, motivating the team, building relationships, controlling expenditures, visualizing the deliverable, and staying focused on results.
- The product/project manager must possess leadership skills to be successful.
- Anyone can develop and improve their leadership skills.
- Identifying stakeholders is important. This will help clarify product features and desires.

Discussion Questions

1. Sometimes, executives do not know their role in projects. Some are completely "hands off" and feel once they provide funding, the project is in the hands of the project team. Others may tend to be overly involved, meaning they are micromanaging the team. What are your thoughts on this? How would you deal with each of these supporting executives?
2. Why is the development of a new product considered a project?
3. When a company is "publicly held," what does this mean?
4. Within your own company, how has Sarbanes-Oxley affected how you plan and execute a project? (This assumes your company is publicly held.)
5. Who should manage a new product's development, an engineering manager, or a marketing manager?
6. Why is Net Present Value important?
7. Why is controlling budget and schedule important?
8. Your company has $10 million to invest in a new product. The marketing and finance groups have identified three strong projects, but you only have enough money to invest in one of them. The following NPVs have been determined. From a purely financial perspective, which one should your company choose?
   Project 1. NPV = $23,000
   Project 2. NPV = 0
   Project 3. NPV = $51,000
9. As the team leader, you must motivate your team. What are some actions you can take to keep team morale high?
10. What are some of the techniques we can use to identify project stakeholders?
11. To be a leader, personal integrity is a necessary ingredient. However, senior executives of Enron, Tyco, and Worldcom landed in prison. How is it that leaders of these companies ended up in this situation? Did they lose their integrity along the way?

Scope Change Request Form

Product: _____   Date: _____

Submitted by: _____

Brief Description of Change and Justification:
(Attach additional pages if necessary)

Impact to budget:
Impact to schedule:
Impact to quality:
Impact to resources:

Approved by _____   Date: _____
            Product Manager (print)

Signature _____

Note: changes above $20,000 require approval by vice president or higher

Figure 4.7   Scope Change Request – Sample. Developed by David Tennant

Discussion Case 4.1 Organizational Conflict – Discussion Questions Answers

1. How would you handle this ongoing, organizational conflict?

   It is a failure of senior management that this organizational conflict has been allowed to continue. Since this is endemic to the organization, the only solution is change at (or from) the top.

2. What are your chances of getting the two managers to work with each other?

   It should be apparent that these two managers dislike each other and will not be warming up to each other anytime soon. However, both will recognize that they must support the project as part of their job responsibilities. It is likely they will provide the needed support, but you will need to manage this situation and possibly act as referee on occasion. It would be best for you to interact and let each provide their deliverables with as little interaction as possible.

3. Since this is an organizational issue, and has happened with other departments, how can we resolve this?

   See the answer to number 1 above. Beyond that, you may need to start up the chain of command (starting with your immediate boss) to move things along.

4. Is this a case of personality conflict?

   Most assuredly. Both managers are strong "driver" personality types.

5. What does your team expect of you, the product manager?

   You are expected to lead the project. This includes motivating your team, resolving conflict, coordinating the various stakeholders, communicating up and down the organization, controlling budgets and schedules, running meetings, and knowing the right questions to ask.

Discussion Questions – Answers

1. Sometimes, executives do not know their role in projects. Some are completely "hands off" and feel once they provide funding, the project is in the hands of the project team. Others may tend to be overly involved, meaning they are micromanaging the team. What are your thoughts on this? How would you deal with each of these supporting executives?

   This can be tricky. A lot depends on the company culture and the individual personality style of the sponsoring executive. Building a relationship with the sponsor will help with this issue. Also, if you have a strong track record of achievement, he or she is likely to give you more freedom of operation. If you are not well known to the sponsor, you may find he or she is more demanding until there is a higher comfort level with your demonstrated abilities and skills. Nonetheless, relationship building and regular "check ins" with the sponsor will be beneficial in dealing with both types.

2. Why is the development of a new product considered a project?

   New product development is a "project" in the following context:

   - *It will have a budget and schedule*
   - *There will be milestones*
   - *Most likely, you will have gate reviews*
   - *The risks (potential problems) associated with the new product must be identified and mitigated*
   - *Quality is paramount to success*

- *You will need to lead a team and be responsible for resources*
- *Supply chain will be involved with any purchasing activities*
- *The scope of the new product development must be defined and controlled.*

3. When a company is "publicly held," what does this mean?

   This means your company is listed on the US Stock exchange. It issues stock to investors who are then legally owners of the company. It is one way that companies can raise money without borrowing the funds. The owners invest cash in the company in exchange for stock (ownership).

4. Within your own company, how has Sarbanes-Oxley affected how you plan and execute a project? (This assumes your company is publicly held.)

   If your company is publicly held, it is required to conduct its financial reporting with more transparency and company operations have stringent controls in place. This affects most functional areas.

5. Who should manage a new product's development, an engineering manager, or a marketing manager?

   This is most likely dependent on the company culture, the type of company, and the type of product being developed. If it is a general merchandise or manufacturing company, it may be a marketing person leading the effort. If a highly technical firm (e.g., Apple, Google, etc.), it may be an engineering person leading the group. In a medical products company, it could be either person.

6. Why is Net Present Value important?

   NPV takes future cash flows (dollars leaving and coming in) over a specific time period and brings the costs back to today's value. It determines the profitability of a project based on the time value of money and a company's required return on investment. It is very useful when comparing multiple projects.

7. Why is controlling budget and schedule important?

   If the costs get out of control, it is possible that your project may be canceled as the profitability has disappeared. Schedule is important and this also represents money. If your schedule starts to slip, then the budget will most likely also slip. Further, schedule delays may give your competitors time to beat you to market.

8. Your company has $10 million to invest in a new product. The marketing and finance groups have identified three strong projects, but you only have enough money to invest in one of them. The following NPVs have been determined. From a purely financial perspective, which one should your company choose?

   Project 1. NPV = $23,000

   Project 2. NPV = 0

   Project 3. NPV = $51,000

   Although each of the three projects above is profitable, the largest NPV offers the best return on investment. Therefore Project 3 should be selected.

9. As the team leader, you must motivate your team. What are some actions you can take to keep team morale high?

   There are a number of activities the project leader can take to build morale:
   - *Take the team to a professional ball game*
   - *Provide high-quality cups, plagues, tee shirts, etc. emblazoned with the company or project logo*

- *Have the sponsoring executive honor those team members who have gone beyond expectations with recognition and/or a spot cash bonus*
- *When a major milestone is reached, celebrate with cash rewards, gift cards, etc.*

10. What are some of the techniques we can use to identify project stakeholders?

- *Determine who is funding the project and ask him or her who are the key players.*
- *Look at the WBS (Work Breakdown Structure) to determine which functional areas and suppliers are involved.*
- *Go to each functional area that you know is engaged in the product and ask questions.*

11. To be a leader, personal integrity is a necessary ingredient. However, senior executives of Enron, Tyco, and Worldcom landed in prison. How is it that leaders of these companies ended up in this situation? Did they lose their integrity along the way?

Several perspectives can be presented for this question.

- *The investigators and prosecutors in these cases felt that the senior executives were greedy. By falsifying the accounting reporting, they were able to keep the stock prices of these companies artificially high. And all were compensated on keeping the stock price high with stock options.*
- *Some would suggest that our economic system (capitalism) has an over-reliance on short term performance. That is, will the company make their quarterly sales and revenue numbers? Will they pay a stock dividend next quarter? Will they beat or miss Wall Street analysts expectations? These types of expectation put pressure on executives to meet numbers (or risk losing their jobs).*
- *To meet shareholder and Wall Street expectations, it is theorized that Enron cooked their books thinking, we'll make up for it with next quarters sales and revenues. When the next quarter's sales miss expectations, the books are cooked again and over time, the numbers fall further and further from reality.*
- *Enron's accounting firm, Arthur Anderson, made more money in consulting to Enron than by auditing their financial statements. Consequently, there was a strong conflict-of-interest in this relationship. The accounting firm worked hard to "justify" the auditing that was performed, even though it was illegal. Arthur Anderson went bankrupt as a result of the investigation; and, Sarbanes-Oxley legislation has clearly stated the required separation of consulting and auditing activities.*

Notes

1 Finzel, H., The Top Ten Mistakes Leaders Make, David C. Cook pub, Colorado Springs, CO, 2007.
2 Maxwell, J., The 21 Indispensable Qualities of a Leader, Thomas Nelson, Inc., pub, Nashville, TN 1999.
3 Ulrich, D., Results-Based Leadership, Harvard Business School Press, 1999.
4 Steven Covey, The Seven Habits of Highly Successful People, Simon and Schuster, New York, NY, 1989.

Bibliography

Garrison, R. and Noreen, E. (1997). *Managerial Accounting*, 8th e. McGraw-Hill.

Jordan, B., Ross, S., and Westerfield, R. (1998). *Fundamentals of Corporate Finance*, 4th e. McGraw-Hill.

Mantel, S. and Meredith, J. (2002). *Project Management, A Managerial Approach*. John Wiley and Sons, pub.

Zenger, J. and Folkman, J. (2002). *The Extraordinary Leader*. McGraw-Hill.

5

Getting Started – Project Approved

Product/Project Management and Engineering

Taking the Business Case from Concept to Reality

As discussed in Chapter 2, the business case is a detailed analysis or feasibility study to develop a new product, merge or acquire another firm, or have the company embark in a new direction. The business case should contain enough information and recommendations that the executive team can make an informed decision. Recall the "typical" Table of Contents:

• Summary and Introduction	• Legal and regulatory issues
• Objectives	• Estimated costs and timeline
• Description of new product/project	• Resources needed
• Why should the company do this (what is driving this effort)?	• Social implications
	• Alternatives considered
• Cost-benefit analysis (ROI, NPV)	• Recommendations

Once a decision has been made to move forward, the contents of the business case must come to life. How will this occur? The "project" leader will pull together the needed resources and bring all the business case contents to fruition. Ideally, the product team will have had some input during the creation of the business case.

However, there are a few red flags to be aware of in the transition.

A business case may go into great detail and analysis regarding scope, costs, schedule, etc., but the real planning will occur once the development project is approved, and the team is mobilized.

The true costs, ROI, logistics and other details are difficult to quantify in the business case. This is because cost estimates or supplier quotes are budgetary, timelines are based on best guess scenarios and resources are always assumed to be available when needed. Consequently, the product's development will have significant costs and risk potential. It will take formal supplier quotes and a full cost analysis of each task or activity associated with the project to obtain the true costs of development.

Like costs, schedules are typically high level, lacking in detail. A real schedule will be dependent on supplier deliveries, acquiring people talent when needed, and considering risks or potential problems.

Product Development: An Engineer's Guide to Business Considerations, Real-World Product Testing, and Launch, First Edition. David V. Tennant.
© 2022 John Wiley & Sons, Inc. Published 2022 by John Wiley & Sons, Inc.

Engineering, Research and Development: What does this mean?

Research and Development (R&D) can apply to a variety of activities. However, there are two types of research: Basic and Applied.

Basic Research

Basic research is focused on expanding knowledge and creating new theories (or modifying existing ones) simply for the sake of knowledge. Basic research strives to study new ideas (sometimes called pure or fundamental research), is curiosity driven and aims to seek new knowledge that can be universally applied. For the most part, it is theoretical. Most basic research is performed by government laboratories, universities, think tanks, and sometimes private industry. Examples of basic research include:

1. CERN is the European Organization for Nuclear Research. Literally, it is derived from the French CERN, an acronym for the French Conseil European pour la Recherche Nucléaire.[1]

   CERN's mission is the study of fundamental particle physics using particle accelerators (faster than the speed of light for less than a second). This is accomplished using a Large Hadron Collider. Specific knowledge is sought regarding the concepts of matter vs. anti-matter, understanding the Higgs-Boson particle (mass) and the discovery of additional physical properties at higher energy levels. The mission of CERN is stated on their website:

   > "At CERN, our work helps to uncover what the universe is made of and how it works. We do this by providing a unique range of particle accelerator facilities to researchers, to advance the boundaries of human knowledge.[2]"

CERN is primarily a European venture funded by 23 "member" European countries. These include France, Germany, the United Kingdom, Spain, Switzerland, and others. The member countries provide the capital (funding) to build, operate, and maintain the facility. Observer status countries (not full members) include the United States, Russia, and Japan. Over 600 universities and institutes participate in experiments at CERN.

2. The Human Genome Project (HGP), an international collaboration that successfully determined, stored, and rendered publicly available the sequences of almost all the genetic content of the chromosomes of the human organism.[3]

When the human genome project started, there were many contributing scientist and researchers from around the world in this international collaboration.

> The HGP, which operated from 1990 to 2003, provided researchers with basic information about the sequences of the 3 billion chemical base pairs (i.e., adenine, thymine, guanine, and cytosine) that make up human genomic DNA (deoxyribonucleic acid). The HGP was further intended to improve the technologies needed to interpret and analyze genomic sequences, to identify all the genes encoded in human DNA, and to address the ethical, legal, and social

implications that might arise from defining the entire human genomic sequence.[4]

The use of sophisticated computers and software allowed the project to proceed quickly. This was directed by the US Department of Energy (DOE) and the National Institutes for Health (NIH).

The previous examples illustrate the type of "basic" research that is performed. The costs of this type of research are high with an unknown end benefit. Consequently, this type of research is performed with government funding or grants. However, it is important to note that results from basic research will many times be used by private companies who can commercialize the results.

Here is a summary of basic research:

- Basic research is sponsored by government agencies or organizations (universities for example) to further general knowledge or scientific principles.
- The results of basic research are in the public domain (no ownership). One exception would be research for national defense.
- Results are typically published in a scientific journals or conference proceedings.
- Basic research relies on the reputations of the researchers.
- Basic research produces findings and conclusions, not suggestions or applications.

Generally, private industry (and publicly held firms) must make a profit to stay in business. Therefore, these companies perform "applied" research or, they will find an application resulting from basic research.

Applied Research

Applied research strives to solve specific problems by providing new solutions to issues affecting a target market, group, or society. Its focus is the pragmatic application of scientific principles to solve everyday problems. Examples of applied research include:

- Developing a new vaccine for a specific disease
- Improving the range and efficiency of electric vehicles using new technology
- Inventing the next generation of cloud-based computers
- Increasing crop production in a small area to feed more people
- The development and testing of a new wing design for general aviation use.

It is appropriate to note that applied research is in the domain of private industry. It is funded by company budgets or from client funding.

Practical applications resulting from basic research may be funded by government grants and private industry. A specific example is the Department of Energy funding clean coal technology. Several of these projects have been funded with mixed success. Government provides the funding and private industry, in this case electric utilities, attempt to find solutions relative to reducing or eliminating air pollution from coal. The cost of failure is very high, and this can be mitigated by government funding (minimizing company financial losses), but the reward can be very high and profitable, if the applied research project succeeds. There is never a *guarantee* that applied research will be successful.

However, if companies and research institutions were risk averse, basic, and applied research would never occur and we would still be riding in horse drawn carriages. Companies cannot be averse to risks: developing new products, services, or ideas. Companies must take risks to succeed, stay ahead of their competitors and remain profitable.

> *This means that applied R&D of new products and services is the future lifeblood of any company. The world does not revolve around engineering, marketing, or technology; it revolves around being profitable.*

The only way to remain competitive is to develop new ideas into products.

Project Management in Product Development

One key failure point in product development is the lack of robust project management. In the author's experience, significant failures are due to poor planning, lack of schedule, budget, or scope control, lack of risk reviews, poor communication, etc. And poor scope definition (related to poor planning) is a major cause. Project management (PM) is the key to coordinated success and a structured process to achieve objectives. In discussions with numerous product development professionals, this is a common consensus: having a robust PM process increases the probability of product development success.

There are numerous books and courses on the topic of project management, and several key concepts have been sprinkled throughout this book (for example, risk reviews in Chapter 8, stakeholder analysis in Chapter 3, etc.) and this chapter will provide more depth on the mechanics of project management. It would be impossible to provide all aspects of project management into this chapter and the reader is encouraged to do further research on this topic. Please note that all projects are temporary. Each project has a definite starting point and a definite ending point. Projects go off track with constant changes to product features (scope creep), lack of budget control, etc. All projects should have clearly defined objectives and a clear vision of what the "deliverable" will look like. Note that the author uses the terms project manager and product manager interchangeably.

First, let's consider, what is the difference between a product and a project.

- Product development is – developing something of value in the marketplace through design, planning, testing, and manufacturing. It is also managing its modifications, marketing, sales, and distribution.
- Project Management is – planning, implementing, and controlling a group of activities to accomplish a desired result.

Examples of projects include:

- Developing the next smart phone release
- Building a new hospital
- Developing the next generation of electric vehicles
- Developing a marketing plan for your company
- Researching a cure for arthritis
- Developing a new product for the marketplace.

Another definition includes that developed by the Project Management Institute® (PMI®). According to PMI®, project management is defined as:

> Project Management is the application of knowledge, skills, tools, and techniques to project activities in order to meet project requirements[4]

Project management includes:

- Balancing the requirements and needs of stakeholders
- Establishing clear objectives
- Monitoring and controlling the competing demands of schedule, budget, scope, and quality
- Managing the stakeholders and project communications
- Anticipating potential problems (risks) and developing mitigating solutions.

The definitions and activities of product and project management are very similar. Therefore, a new product's development should be considered and managed as a project.

PM Truisms:

- Projects solve problems (developing new products, earning a return on investment, developing solutions to specific problems).
- The project leader/manager is ultimately responsible for the success (or failure) of the project. This applies equally to the Product Manager.
- The PM is generally authorized to obtain and expend company resources: money, people, equipment, etc.
- No two projects are identical.
- Product/project management usually requires the engagement of different functional areas and outside suppliers.
- The product manager needs project management skills to be effective and successful.

Why Do Projects Fail?

We could easily provide a list of 20 or 30 reasons why projects fail. However, Table 5.1 Reasons Projects Fail, is the author's Top 10 list:

Table 5.1 Reasons Projects Fail.

	Failure Item	Comments
1	Poor planning and Execution	Requirements not confirmed, lack of adequate cost estimating, failure to anticipate problems, lack of clear roles, etc. Poor planning will also lead to poor execution.
2	Inadequate resources	This is related to item 1. Resources can refer to people, equipment, software, dollars, etc. If the project has robust planning, this problem should not occur.
3	Poor communication	This can be a combination of a company being "siloed" and/or the lack of the project leader in communicating effectively with stakeholders. This can also lead to departmental conflict.
4	Objectives not clear	Failure to reach objectives by consensus will lead to conflict and potential failure. The objectives should be reviewed on a regular basis to confirm they are still valid in the marketplace.
5	Poor leadership	It takes a special person, with excellent communication skills, to lead, motivate, and drive a project to completion.
6	Continuous scope changes	Commonly called "scope creep" it is the continuous growth of the project's objectives and tasks without a formal review process. This inevitably leads to budget overruns, schedule slippage, and conflict. Even using Agile, some form of scope change review is needed.
7	Failure to heed warning signs	Projects will always give warning signs when they are heading for the ditch. These include budget overruns, missing schedule milestones, and the worst sign: people are bailing out of your project. Project trends develop 20% into a project. If one can pay attention to warning signs, you may have up to 80% of the project to take corrective action.
8	Unrealistic expectations	For example, the PM is sometimes given a budget and deadline to develop a new product before detailed planning has begun. This is unrealistic and unlikely to have a good outcome.
9	Lack of risk reviews and risk management	One of the biggest tools to keep projects humming along is the use of project risk reviews. This is a structured process to anticipate future problems and develop solutions to prevent or minimize their occurrence. This topic is covered in detail in a subsequent chapter.
10	Poorly defined roles and responsibilities	It is important that each member on the product development team understand his or her role and responsibility. Lack of this understanding many times leads to duplication of effort and organizational conflict.

Table by David Tennant.

The reader will note that the above problems are not technical. Technical problems do occur, but they can usually be solved with time and talent. However, the problems above are managerial which can be more difficult to resolve. Projects generally fail due to managerial issues more than technical issues. This raises the question: does the product or project manager need technical skills or managerial skills?

What should be in a project plan?

The project plan is a guiding document that is the roadmap to project success and should be developed by consensus. The project plan is the mechanism to translate the business case into an action plan. This is illustrated in Figure 5.1 From Concept to Action.

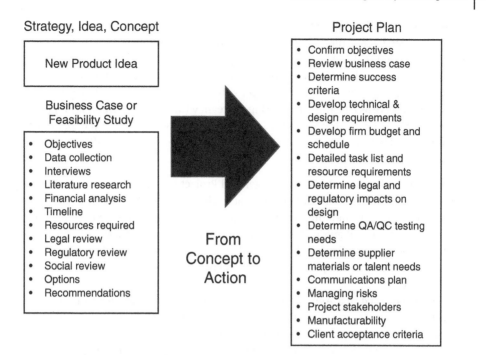

Strategy, Idea, Concept

New Product Idea

Business Case or
Feasibility Study

- Objectives
- Data collection
- Interviews
- Literature research
- Financial analysis
- Timeline
- Resources required
- Legal review
- Regulatory review
- Social review
- Options
- Recommendations

From
Concept to
Action

Project Plan

- Confirm objectives
- Review business case
- Determine success criteria
- Develop technical & design requirements
- Develop firm budget and schedule
- Detailed task list and resource requirements
- Determine legal and regulatory impacts on design
- Determine QA/QC testing needs
- Determine supplier materials or talent needs
- Communications plan
- Managing risks
- Project stakeholders
- Manufacturability
- Client acceptance criteria

Figure 5.1 Concept to Action. Figure by David Tennant

Traditional and Agile Project Management

Traditional project management has a project leader that directs the activities and resources of the company in planning and executing a project. It is focused on defining and controlling the scope of a project from beginning to end. Upfront planning is robust and documentation (status reporting, communications planning, risks, etc.) is considered necessary. This type of management is appropriate for projects such as building a manufacturing plant, constructing a hospital, or a road infrastructure project. And there are occasions when new product development is long-term. The focus is to provide a deliverable within schedule and budget with minimal changes (disruptions) and with the agreed-to quality (functionality, meeting standards and client requirements).

On the other hand, Agile project management has as a similarity in that the deliverable must also meet quality requirements. But the similarity ends there. Agile PM is focused on a "product owner" providing an overall vision and facilitating the development team's activities. Agile focuses on robust planning for the immediate tasks while leaving detailed planning for future activities, well, in the future. Documentation should be brief (the bare essentials) and the deliverable is developed as an iterative process. Planning is also iterative, developing along with the product or project. Changes (i.e., disruptions) are welcome and the client has input at every step of the product's development. Key points in an agile project:

- It is difficult to anticipate every design or feature of a proposed product at the very beginning. It is much easier (and less costly) to accommodate changes as the process develops rather than later.

- The deliverable should meet the bare minimum requirements. Future requirements can be added later as the product unfolds; or saved for version 2.0.
- The team is left to develop the product/project as needed. Capable people are given the freedom to apply their specialized skills and talents to their activities.
- The Product Manager or Project Manager will have less of an authoritative role and will rather serve as a facilitator: assisting the team with resources, training, supplies, coordination, and communications. This is a very different role from the traditional PM that wielded significant influence and control. Some PMs have difficulty transitioning to this "new" style.

While the author uses Product Manager and Project Manager interchangeably, in some companies there is a distinct difference. This is especially true in IT projects. Some firms will consider the Product Manager as one who determines customer needs, provides overall guidance and communications, and interacts with the client on a solution or needs basis. There will also be an assigned project manager who focuses on the day-to-day project activities such as budgets, schedules, risks, team activities etc.

It should be noted that Agile project management came to fruition within the software/IT industry. The success rate of IT projects; that is, meeting budget, schedule, and functional requirements was abysmal. Something better had to be devised for these types of products. While Agile started as an IT "system" of management, Agile has been transitioning to other functional areas such as engineering and manufacturing. In some cases, companies have been using a hybrid model, combining the best of both systems where appropriate.

Because both Agile and Traditional PM are relevant, this chapter will provide information and illustrations on each. It should also be stated that Agile is not a "one project" only process. Companies that decide to embrace Agile must be fully supportive in how Agile truly functions. It is most likely a different approach from what has occurred in the past. Without company support for how Agile works, it will be very difficult for the product team to be successful.

Remember, the desired result of each project management system is the same: a successful and profitable product. The difference is in the approach.

Traditional Project Management

The project leader shepherds the project plan's development by coordinating with all stakeholders and the business case.

Below is a suggested project plan Table of Contents. Each topic is discussed in detail further below.

- Executive Summary
- Key Stakeholders
- Scope Statement
  - Objectives
  - Assumptions and Constraints
  - Critical Success Factors
  - Major Milestones
- Summary Tasks and Budget Discussion
- Resource Requirements

- Risk Management
- Project Communications
- Managing Changes in Scope
- Quality Assurance and Quality Control
- Supply Chain (purchasing, expediting, and contracts)
- Project completion – recognizing success
- Appendix
  - High-level Schedule and Budget
  - Tools (forms, templates, etc.).

Sample Project Plan – Detailed Table of Contents

(Developed by David Tennant)

Executive Summary

This should provide a high-level review of the plan's contents: key objectives, budget, timeline, and what the deliverable(s) consist of. If a business case was the basis of the approved project, this should be referenced. It may also be appropriate to provide a project organization chart.

Key Stakeholders

The major stakeholders should be listed including their role. For example

- VP Marketing – Provides funding and high-level direction
- Director Marketing – Project leadership and direction, day-to-day responsibility
- Director Engineering – Technical expert, directs engineering efforts
- Supplier XYZ – Providing critical components for new product
- Director Supply Chain – Point of contact for all procurement items and suppliers
- Superintendent Production – Lead point of contact for all manufacturing tasks.

Scope Statement

Exactly what are we striving to accomplish? The following items will layout the scope (i.e., breadth) of the project:

Project Justification – (From business case) Objectives

Generally, more than one objective, deliverables, dates, etc. For example:

The deliverable will be a (item description) with a target retail price of $_____ and a projected development and manufactured cost of $_____. Product will be offered to the public by (date). The first production run will consist of _____ (number of units).

- Assumptions and Constraints – Examples might include:
  - The budget and schedule are realistic (assumption).
  - Management supports this new product's development (assumption).
  - All production raw materials will arrive on time (assumption).
  - Safety during product testing will be a priority (constraint).
  - Funding is tight, budget must be adhered to (constraint).
  - Production must start 3rd quarter of next year (constraint).
- Critical Success Factors – these represent things that must occur. Lack of these could be a showstopper. Examples may include:
  - Production facility cannot shut down due to problems or scheduled maintenance. This could delay manufacture of new product (meaning late to market).

- Supply chain must ensure timely delivery of control chips which are a (worldwide chip shortage?) key component in new product.
- Lead R&D Specialists (experts): we cannot afford to lose subject matter experts (SMEs). SMEs leaving the company could impact schedule and create a knowledge gap.
- Major Milestones – Key dates in the project lifecycle. For example:
  - June 1 – Engineering design starts
  - August 1 – RFP Issued to potential suppliers
  - September 15 – Contract award to successful bidder
  - November 3 – Gate review
  - December 10 – Engineering Design completed
  - January 20 – Prototype testing begins
  - February 25 – Gate review and test results
  - March 1 – Full scale production begins

Summary Tasks and Budget Description
This section should describe the major tasks (design, prototype testing, data review, design or production revisions, distribution of product, etc.). The budget should describe the costs of major items, design or specification changes, manufacturing, and distribution costs. The budget should also describe the costs of all tasks (people, equipment, external support such as suppliers, consultants, etc.)

Resource Requirements
The discussion should include, based on tasks and schedule events, what type of people expertise is needed, equipment, hardware, software, training, etc. The skilled craftspeople (fabricators, welders, programmers) should also be indicated.

Risk Management
A description of how often risk reviews will occur and type of review. A risk is a potential future problem. How will we anticipate these? Development of strategies in advance that will minimize or prevent risks.

Communications
This section should describe what communications each stakeholder (internal and external) should receive. Status reports, risk reviews, contracts, purchase orders, etc.

   Note that stakeholder correspondence should be targeted. Everyone does not need to receive copies of all correspondence.

Managing Changes to Project Scope
How will scope changes be managed? What is the review process? Who has approval authority and at what level (that is, dollar amount)?

Quality Assurance and Quality Control
How will we ensure quality of design and manufacturing? What testing and data collection is required? How will deficiencies be handled? How will continuous improvement occur? What is an acceptable rejection or error rate?

Procurement Plan
Outline the Supply Chain responsibilities. Which external suppliers are required? What type of contracts are appropriate? Do we have an approved supplier list? Who will be responsible for contract administration?

Project Closing
How will we define success? Under what criteria will the client accept the deliverable (product)? When should contracts and accounts be closed? Will a lessons-learned review be necessary?

The project plan may mean different things to different people. For example, in the Information Technology (IT) field, a project plan may mean a very detailed schedule.

In other disciplines, the project plan includes all the discussion points covered previously, including a detailed schedule.

For IT practitioners, a project charter is a document similar to what we have been describing as the project plan. It is up to the project leader to clearly communicate to all parties what is to be included in the project plan or charter. It also helps to have key stakeholders, including the client (whether internal or external) sign off on the plan. This would also include the executive sponsor.

As previously noted, traditional PM is suited for long-term well-defined capital expense projects.

A True Case Study - Company Dysfunction and a Lack of Project Management

Company X had a multi-functional team pulled together to support new product development. The team consisted of people from Engineering, Marketing, Finance, and Manufacturing. The Field Service group was not always involved in initial team activities, even though this is where the product had to be installed and troubles fixed during client installations.

It was found that many client start-up issues could have been prevented if the field services group was involved in product development in the early stages. Once a client implementation ran into problems, there was finger pointing and blame games between the product team members, field services, and corporate. Also, in the case of Company X, the corporate bureaucracy became an obstacle to efficiently address product introduction problems. For this company, multiple failures led to $300 million of scrapped materials and hardware. The company, a $19 billion corporation, eventually went bankrupt, in part, due to their inability to develop reliable products as quickly as their competitors.

Specifically, a new laser printer was developed for commercial businesses. The marketing and sales group offered an inexpensive maintenance plan to their customers. It was found the plan was profitable for small businesses. However, large businesses were printing thousands of pages per day, which led to more frequent maintenance calls and this segment of customers was causing greater costs thereby making the maintenance plan unprofitable. The field services group devised a multi-level plan for small, medium, and large business customers, thereby restoring profitability. This confirms the importance of having a wide perspective on the team and to include the end user of the product. In addition to the product, this is also a service issue.

Observations and Lessons Learned

It was generally a marketing person who served as the team leader or product manager. The Marketing group was generally the sponsor and provided funding for new product development. The engineering team led the effort during design phase; then production would lead the effort during the production or manufacturing phase. That is, each member of the team would lead the project when it was their lead role for the product. Each department took "star billing" as needed.

There was a formal product development process whereby anyone could suggest changes or modifications to a proposed product's features or scope. A business plan was created and sent to senior management for approval. Potential financial performance was always high on the list of product development.

In the early stages, estimates included expected pay scales and a corporate budget for initial costs. A detailed cost estimate was refined as the project moved forward and costs become clearer. The financial model would include the costs of each group (marketing, production, supply chain, etc.). A detailed proforma forced the contributing groups to document what they would be doing. This is in addition to a listing of roles and responsibilities.

The overall business plan, appropriately detailed, forced transparency in product development. This reduced finger pointing and blaming and instead changed to a search for solutions. The PM played the role of facilitator. Additionally, the PM had a strong relationship with the sponsor – usually a high-level executive – so that assistance with issues could be obtained quickly when needed.

As with any publicly held company, there will always be pressure to increase sales and bring new products to market faster. The Finance group was always exerting pressure to meet the next quarters "numbers." This led to new products being pushed out the door before they were ready. While the sales numbers looked good, the headaches and excessive costs created by warranty and client issues were financially detrimental.

Eventually, a budget planning model was developed to predict service failures (and thereby, costs associated with failures) to improve product launching. This included part failure rates. This could then be translated into dollar costs.

Team meetings occurred on a monthly and sometimes weekly basis to effectively coordinate efforts. The marketing group performed program risk reviews.

The field services group considered Company X to be an innovative and creative company – far ahead of their competitors. However, the corporate bureaucracy, poor organization, and senior leadership led to corporate dysfunction.

The Takeaways

- Strong project management skills and knowledge are needed for successful product development.
- Organizational issues can derail new product development and, if severe enough, can bankrupt companies.
- New products should be developed from the end user's perspective.
- The financial model needs to consider the cost of part failure rates.
- In the case of software, the training department should also be on the team (great end-user perspective).
- Supply chain: can offer insight into global logistics.
- Administration: can offer people/HR resources.
- Sales can add value in clarifying customer perspectives and wants.

Table 5.2 Roles of the Project Manger Vs. Project Engineer.

Project Manager	Project Engineer
• Overall project responsibility	• Directs the technical efforts of the engineering team
• Coordinates functional groups	• Coordinates with Project Manager in working with other functional areas
• Leads project communications	
• Manages scope, budget, and schedule	• Ensures that quality is a priority
	• Works with suppliers on specifications and technical issues
• Serves as liaison between suppliers, senior management, project team and others	• Coordinates, when appropriate, with manufacturing
• Monitors progress and provides status updates	• Oversees design, modeling, and development of prototype(s)
• Manages stakeholders	• Interprets data from prototype testing
• Facilitates risk reviews	• Revises design as needed

Table developed by David Tennant (author)

How is a project manager different from a project engineer? There is some confusion in the difference between a project engineer and project manager. The roles are similar, but distinct. Table 5.2 provides the distinction between a project manager and a project engineer.

Developing and Controlling Scope – Using a Work Breakdown Structure (WBS)

Developing the Work Breakdown Structure is an integral part of Product/Project planning. The WBS is simply a list of all the tasks and activities that must occur for the project to be completed. Consequently, the WBS defines the scope of the project.

Figure 5.2 Air Emissions Monitor WBS, illustrates how the tasks for a project can be arranged in a chart form. The WBS attempts to arrange all the known tasks or activities that are required to complete the new product's development.

Clearly, a project of this size would include many more tasks, but this figure illustrates the approach and how the WBS forces the product team to think through the process.

The WBS is a useful tool:

- For each block on the diagram, a person or group can be assigned responsibility and accountability.
- A cost for each block can be derived. The costs can then be summed to obtain an overall project cost estimate or budget.
- Each activity (block) can be placed into a project schedule.
- All the blocks together represent the full scope of the project (traditional PM).
- The WBS is developed by the team and shows functional interaction.

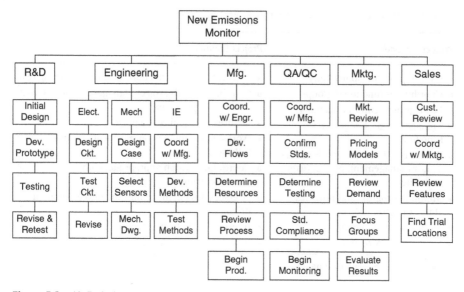

Figure 5.2 Air Emissions Monitor WBS. Figure by David Tennant

Developing a Budget and Cash Flows

Chapter 4 offers further high-level discussions on budgets and cash flows. It is worthy to note that some development projects may span across two budget cycles. For example, a development project starts in August and will finish in April of the next year. Since many companies' budget for each fiscal year, it is important that adequate dollars are apportioned over both fiscal years budgets.

Refer to Figure 5.2 which shows a WBS for a new air emissions monitoring device. If we determine a cost for each task or block on this diagram, we can them sum them together for each category as shown in Figure 5.3 Air Emissions Monitor Cost Summary.

As shown, the total product development cost, or budget, is $8,275,000. To reinforce a previous point, the market will determine the selling price. Based on market research, competitor evaluations, and other factors, your marketing team is projecting that the retail price of each monitor will be $110,000. This means that approximately 75 units must be sold to break even. That is a projection, but will the market support that price?

Your product or marketing team can suggest a price, but all that you can really control is the cost of development and production. Also, to have a true reflection of total costs, it would be appropriate to include a dollar amount for potential warranty repair work, which would drive the overall budget, and breakeven point, upwards. The WBS is a strong tool that allows the team to review resource requirements and develop an estimated cost. From this, a cash flow diagram and profitability analysis can be performed.

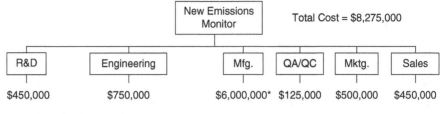

• Mfg. Includes first 100 unites at $60,000 each

Figure 5.3 Air Emissions Monitor Cost Summary. Figure by David Tennant

Using the WBS to Derive a Schedule

Similar to budgets, each block or task on the WBS can be placed into a schedule. In essence, the task list in a schedule is a WBS.

Comparison of a Traditional vs Agile Project

The following is a hypothetical scenario of one development project. It is a comparison of an Agile vs. Traditional PM approach. Let's consider the development of a new technical product: an air emissions monitor as previously discussed. An air monitor is used by industries to monitor air emissions from a factory or combustion process. For example, a paper-pulp plant uses a boiler to produce steam for use in making boxes, paper, and other products. To stay within compliance of environmental laws, the emissions from the boiler (usually combusting wood, or waste wood by-products, or other fuels) is monitored to help control emissions into the atmosphere. The monitor can assist the plant in controlling the combustion process so that it is efficient and clean.

Refer to Figure 5.2 Air Emissions Monitor WBS. This defines the scope of our product's development. Our first attempt to develop this product will use the traditional PM approach. Reference Figure 5.4 for a traditional PM approach.

Business Case

The business case is a feasibility study to provide a concept with supporting content. The content may include costs, ROI, benefits, preliminary timeline, budget, and needed resources. It usually presents several options along with a recommendation. The purpose of the business case is to help senior management in making a decision. The business case, once approved, sets in motion the beginning of our project, or rather, our product's development.

Note that the business case is prior to project initiation. The project is not authorized and therefore does not start until the business case is approved and funds are allocated. The reader will note that the phases of a project may have overlap, but not always. In the traditional approach, the phases are generally preferred to be separate, that is, sequential.

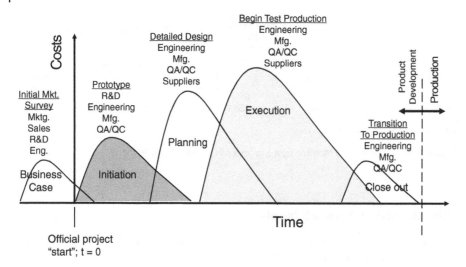

Figure 5.4 Traditional Product Development Project. Figure by David Tennant

Initiation

Once the new product (or project) is approved, there are several activities that occur in initiation. Be aware that companies may have their own approach to initiation, planning, etc., but this section provides the basic concepts.

Many companies use a matrixed approach for staffing projects. For example, the project team may consist of people from different functional areas: engineering, accounting, marketing, etc. This team will most likely be assembled during the initiation phase and the participants may or may not have worked with one another on past projects. A project leader will generally be appointed with some authority to assemble the team and expend company funds. Here are the key activities during the initiation phase:

- Project Leader/Product Leader is appointed by senior management.
- Executive sponsor assigned.
- The project team is assembled.
- The team reviews the business case and confirms the objectives.
- Key stakeholders are identified.
- A project charter is issued. Not all companies issue a charter, however, this is a recommended practice. The charter should list the objectives, indicate a rough budget and schedule, formally recognize the product leader, and list the executive sponsor.

Recall that the traditional PM approach incurs significant upfront planning, changes are considered a negative occurrence, and customer input beyond initiation is accepted, but limited.

Planning

Although a rough schedule and budget may have been determined in the business case, this must be verified through extensive planning and determining the scope of the product's development. The planning phase will build upon the work done in initiation and the business case. Typical activities in planning include:

- Building a Work Breakdown Structure. This defines in detail the scope of the project. It also brings the team together to build the scope by consensus.
- Determine the project costs or budget (using the WBS).
- Derive the project schedule (using the WBS).
- Assign responsibilities for project activities.
- During status meetings, the finance person will provide updates on spending, expected margins, and product net present value. The Product/Project Manager is responsible for staying within budget.
- The project plan is developed by consensus and details how scope changes will be managed, types of communications, roles and responsibilities of project participants, etc.
- Perform a risk review: what potential problems lie ahead and how can we prevent them? Note: it is customary to perform multiple risk reviews through a project.
- Develop product specifications, parts lists, and materials needed.
- Working with Supply Chain, issue RFPs and obtain quotes from qualified suppliers.
- Award contracts to suppliers and consultants.
- Issue purchase orders for needed materials for full manufacturing.
- Determine which suppliers should be on the core team.
- Work with production to plan factory flows, equipment needed (robotics, numeric control machines, etc.).
- There will be design review meetings and product status meetings throughout the project's duration.

At the end of planning, it is likely that some team members, their participation complete, will be released back to their functional areas or assigned to another project.

It should be apparent that significant work occurs during the planning phase. Depending on the type of product under development, the planning cycle could take a year or longer. At some point, the design will be frozen and scope changes will be discouraged.

Once planning is completed, the project now moves into the execution phase.

Execution

For our new monitor, this is where production of the new product begins. It may initially be a trial run of say five or ten units to confirm that manufacturing is set up correctly, specifications and quality standards are being met, and the units are functioning properly. Activities in this phase may include:

- Develop prototype of new monitor and put through testing. The results of testing data and QA/QC will determine if engineering design is adequate or needs changes.
- It is likely that one or two end-users (clients) will be selected to try the prototype model under realistic conditions. This is critical as the opportunity to obtain user input, likes, dislikes, ease of use, etc. is obtained. This provides valuable feedback to the design team prior to full scale production.
- Once the design is finalized and the product approved, production drawings will be released to manufacturing.

- Coordinate with manufacturing to assist with new workflows, plant operations, troubleshooting, new assembly techniques and worker training.
- Ensure that proper materials and quantities are in warehouse.
- Plant machine operators and craft labor are trained and ready to proceed.
- QA and QC processes are in place.
- Full scale production commences.
- Finished monitors are packaged and shipped.

This assumes that the sales team has been obtaining orders and marketing is promoting the product. Client acceptance in the marketplace signals that the project is completed.

Close Out

While product acceptance in the marketplace means our project is complete, there are several activities that must occur:

- The project accounts are closed (new production accounts will be set up for manufacturing).
- Contracts are closed.
- A "lessons learned" meeting and report may be produced.
- Remaining team members will be reassigned.
- Project documentation is filed (contracts, drawings, etc.).
- The project is formally closed.

Agile PM

Agile has found acceptance primarily in the software development industry. However, some aspects of Agile can be adapted to product development. The Agile Manifesto list four key tenets that are different from traditional PM. These are presented as listed by the authors:

Manifesto for Agile Software Development[5]

We are uncovering better ways of developing software by doing it and helping others do it. Through this work we have come to value:

- Individuals and interactions over processes and tools
- Working software over comprehensive documentation
- Customer collaboration over contract negotiation
- Responding to change over following a plan.

> While there is value in the items on
> the right, Items on the left are valued more.

The reader is encouraged to learn about the principles of Agile project management through the various publications that offer more depth. However, it is worth noting that Agile is focused on teamwork, becoming more efficient, involving customer input,

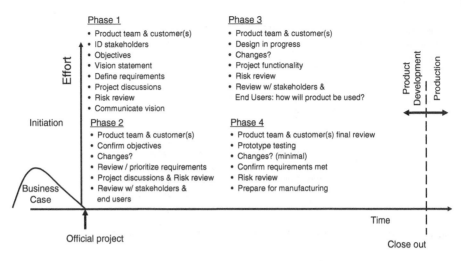

Figure 5.5 Agile Hybrid Product Development Project. Figure by David Tennant

and providing team members with the support and trust they require to be successful. Agile focuses on people, communications, product, and flexibility.[5]

We will now consider developing our air emissions product using a hybrid Traditional-Agile approach.

It should be noted that on products involving engineering or manufacturing, it will be possible to utilize only parts of the Agile concepts. This is because manufacturing may require retooling and training of workers on new machinery; and some product components may have long lead-times from time of order to delivery. Agile cannot necessarily speed up delivery of parts from suppliers.

And finally, clients may not want to participate in product development – that is what they are paying your firm to do. Figure 5.5 Agile Hybrid Product Development Project illustrates the use of both traditional and Agile components in product development.

The Vision Statement

The vision statement is a useful tool to begin any project whether Agile or Traditional. Figure 5.6, Developing the Vision Statement is a process approach to build consensus on this topic.

It should be as concise and specific as possible.

Air Emissions Monitor Vision Statement

The air emissions monitor is being developed for the utility, industrial, and wood-paper products industries. The monitor is a highly complex instrument measuring air emissions to help our customers monitor their air quality and remain in compliance with environmental regulations. Its new technology will increase accuracy, be cost competitive, reduce maintenance costs, and connect with users using wireless technology.

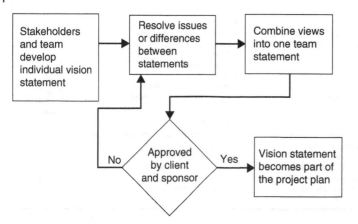

Figure 5.6 Developing the Vision Statement. Figure developed by David Tennant

Table 5.3 Roles Differences of Traditional Vs. Agile PM..

Traditional PM	Agile PM[5]
• PM directs team and project	• Self-organizing team
• Change control through process review	• Embrace change
• Value plans and execution of plans	• Value interaction
• Respond to change after rigorous	• Respond quickly to change
• Review	• Customer collaboration
• Supplier collaboration	• PM is a servant leader and facilitator
• PM is in charge	• Early and continuous delivery of S/W
• Deliverable is predictable at initiation	• Face-to-face conversation is primary
• Formal conversation is primary	• Businesspeople and developers work is priority
• Businesspeople and developers work together daily	• Team should be co-located

Table by David Tennant

This is a concise statement of the product team's *purpose* and should assist in focusing their efforts. The term "purpose" has a more forceful connotation, and some companies may use the term "Statement of Purpose" in lieu of "Vision Statement."

It should be apparent in each phase (Figure 5.5), there is review and discussion to incorporate any changes or ideas that may be generated by the team. In each step, there will be a review of functionality and to design/build a unit with basic required features, not gold plating. Table 5.3 highlights some of the differences between traditional and agile project management. Below are a few issues to be aware of:

- At some point, the design must be "frozen" as continuous scope or design changes will impact cost and launch schedule (i.e., product rollout).
- In companies where a matrixed environment is used, people may drift out of the project, moving on to new assignments.
- Most products are not built as modules like software, so Agile will have limited application.

- Most Agile software teams are co-located and continue with the same people through succeeding release versions. This is advantageous as it builds up expertise and product familiarity. In matrixed companies, each product revision may have a different team thereby losing team cohesion.
- With the global economy, the team may be geographically dispersed.

Agile Hybrid in Action – Marketing Natural Gas in the Southeastern United States: Gas South, A True Story

Natural gas is a commodity product and unregulated in the Southeastern USA. For example, in Georgia residential, commercial, and industrial customers can choose their natural gas supplier. Natural gas is used for heating, cooking, industrial purposes, and for generating electricity.

Gas marketers operate in a fiercely competitive market. Some suppliers attract customers simply by offering low-cost service (some serve at a loss or breakeven point) to win market share. There will always be a segment of customers that choose based on price alone: the lowest cost per therm will always attract these customers.

Gas South, one of the largest gas suppliers in the Southeast, has made great strides in attracting and retaining customers by offering customers what they want, not simply low rates. How does this happen? The firm's marketing group is keenly aware of their customers' needs and challenges. Further, there is a desire to always be ahead of competitors whether in strategic partnerships or offering a variety of gas service rates. The firm decided to use an Agile-traditional hybrid approach to developing new services.

The criteria in developing new services were based on the following:

- Develop a vision statement.
- Use a team approach to developing new services.
- Added customer value: customer preferences.
- Develop using the minimum viable product (MVP) approach.
- Minimal documentation.
- Contain a high level of customer input.
- Quick to market using limited resources.
- Desire for innovative thinking.
- Perform a business case.

The team was cross-functional with individuals who were willing to think outside of the box.

Obtaining Customer Input

Every customer, or potential customer, who calls the service phone number is given the opportunity to complete a short survey. Not everyone responds, but enough feedback is obtained on how well the company is performing.

Every year, a sample of customers and non-customers receive an in-depth annual survey. The survey covers buyer behavior, awareness, perceptions, and customer value drivers. Based on the results of these subsets, further details are pursued using focus groups. This allows the gas company to float new ideas, advertising tag lines, trial messaging, green energy issues, and potential new rate scenarios.

Further, the focus groups represent different customer segments and offers customers the opportunity to discuss what they do or do not like. One issue is that customers don't always know what they want.

Customers can say what they want in focus groups or in surveys but the only way you really know if the solution will work is to test it in the market with real customer behavior. A/B Digital marketing allows you to do this quickly with low cost.

Figure 5.7 Development of Digital Marketing Program illustrates the process Gas South used to design a digital marketing strategy.

One focus group issue that stood out was that most customers do not pay attention to therm usage or pricing per therm details – it was too complicated for most people to understand.

As a result, a OnePrice Plan™ rate product was developed in which the customer did not pay based on how much they used, but rather paid the same amount each month. This offered a convenient way to budget their monthly gas payment and helped simplify the natural gas market. Initially, the plan was offered only to new prospective customers through the company's call center. This allowed Gas South to test this product without significant investment or inclusion in its web-based rate schedules. Using this approach (a minimum product development), it was possible to assess consumer interest and demand, allowing changes or adjustments before rolling out to existing customers. Once real-time data and feedback was obtained, it was rolled out through traditional channels to the whole customer base. It was justifiable to invest more funds in online account services.

The focus groups' feedback resulted in trial limited-service offerings. These could be labeled as MVP: minimal viable product. MVP is a concept promoted by Eric Ries in the book "The Lean Startup." Essentially, a prototype product is developed, typically software-based, with very limited features and minimum effort. These services are evaluated as a prototype by customers allowing one to observe their reactions, behaviors, likes, dislikes, etc. The MVP can be tweaked along the way before initial release. Therefore, the product is adapted to customer preferences during development. More far-reaching changes or features can be added in the next version MVP2, MVP3, etc.

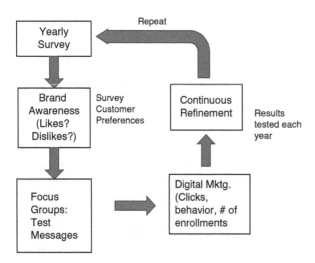

Figure 5.7 Development of New Services Flowchart.

As a result of this approach, the company has been able to grow market share, maintain high customer service ratings, roll out new rate designs quickly, and cultivate an excellent reputation.

Discussion Case 5.1 – A True Story: Product Development Without Project Management

A global telecom company was developing new products for the world market. To meet *anticipated* demand, new capital equipment would be required, and production facilities would be expanded. Marketing had done significant research on customer preferences, R&D was working to perfect the product, and the engineering group was preparing the facility expansion. There were several consulting engineering companies working on facility design and a large construction company was performing the work.

The program consisted of 12 projects and was scheduled for three years duration at a cost of $800 million dollars. It should be noted that the company had its team resources scattered geographically. R&D, Engineering, marketing, and manufacturing (plus consulting engineers and the construction firm) were at the production site; administration and supply chain in another location, and the corporate headquarters in a third location; all of them in different states.

There was considerable pressure to meet the product launch date. The project was behind schedule, and consequently the engineering and construction groups were also behind. The new facilities and equipment were top of the line with considerable redundancy (back-up production) provided.

After one year, the project was six months behind schedule and $70 million over budget. The CFOs office was getting concerned as well as senior management. The product team consisted of extremely bright people, yet the project was floundering. What was the root cause of the project's problems? There were multiple problems, that combined, created the stage for failure.

Because the project was so far over budget, there was concern the new products would be unprofitable, taking years to recover the invested costs. The financials had truly gone into the ditch. R&D was trying to shorten development time; however, short cuts were creating prototype problems. The perfecting of the products had run into problems both technical and managerial. Because of the delays, there was blame and finger pointing. The facility design was dependent on final product configuration (which impacted plant layout), so there was a snowball effect: a delay in one area created downstream delays in others.

There was very little project management processes or skills in place. The people leading the new product effort were highly technical and did not possess the requisite managerial or PM skills.

It has been the author's experience that highly technical people many times (but not always) do not possess effective managerial skills. In fact, technically oriented people sometimes are attracted to engineering and science to minimize management or people issues. Nonetheless, the wrong people were leading this product effort and failed to pay attention to the following:

1. Scope increases and design changes were common, adding to cost and schedule slippage. Some team members were looking to bail out of the project (did they see the iceberg on the horizon?).

2. The new equipment and facilities were over-designed. There was significant redundancy and "gold-plating" which was unnecessary and added to costs.
3. Formal risk reviews were nonexistent. These would have flagged many of the problems earlier in the project(s).
4. The leaders in charge were not strong on communications or conflict resolution. As a result, poor communications led to departmental conflict.
5. While schedules and budgets were in place, there was very little attention or diligence paid to them.
6. No one was held accountable for being late or over-spending.
7. More than one person was leading the effort: there were three project leaders.
8. There was not an overall written program plan to coordinate or tie together the 12 projects.
9. Status reports were non-existent. They were provided only "on request" and were not effective in providing a true picture or in raising red flags.
10. The company had a weak matrix organization, i.e., talented people were "borrowed" from various functional areas. This made it hard to hold team members accountable.
11. Roles and responsibilities were not clearly defined (leading to further conflict).

As listed, there was not any one problem that was contributing to failure, but almost a dozen. The reader will also notice that all the above problems are managerial. It is recognized that some technical problems were delaying the project, but this was being compounded by a lack of project management skills and the inability to focus on the root cause of the problems. This case clearly shows how product development can fail without strong project management processes and talent. This program was very complex where the focus was on technical excellence, but managerial excellence was not in the picture.

Discussion Case 5.2 - Questions

1. You are brought in as a consultant to turn around this failing program. What would be some of your first actions?
2. The three project leaders (Director-level employees) are firmly entrenched in this program. They will most likely resent your presence. How would you deal with this?
3. Would you bring in any outside talent to assist you? If so, who or what kind?
4. Once you have a clear picture of the program's true status, what would be your next steps?
5. How would you propose to get back on track?
6. What are your chances of success?
7. What are some techniques you could use to improve communications?
8. How will you measure success and/or progress?
9. Trust is very important in any consulting engagement. What ideas can you implement to promote yourself as trustworthy?
10. While there are many project management (or lack of such) issues related to this program, there are also many organizational problems. What can you do to help the organization?

This case is clearly a large-scale, Fortune 100 company effort. Think about small firms or even a startup company. Would project management skills help a smaller firm? There are many Fortune 100 companies that have excellent technical and managerial talent, but any of them can fail (or even go into bankruptcy) if not diligent and well versed in project management.

Chapter Key Points

- The business case or feasibility study is an initial review of a potential new product or market. Its purpose is to provide reliable information so executives can make an informed decision.
- Project management skills and leadership are necessary for successful product development.
- Project management proficiency enables the product team to transfer the business case into a plan of action.
- A project management process will lead the team in planning budgets, schedules, communications, risk reviews, and quality assurance/quality control.
- Most companies perform applied research which looks for solutions to specific problems or issues. Basic research is typically performed by universities and government laboratories and is very open-ended.
- Most product/project failures are not due to technical issues, but usually managerial issues.
- Agile project management is more suited to the software development industry.
- It is always necessary to have customer input in developing a successful product.
- The WBS defines the scope of the project.
- A vision statement is always useful in starting a project.

Chapter Discussion Questions

1. Why is the WBS important?
2. Agile has become very popular in the software development industry. Will this displace traditional project management? Provide reasons to support your answer.
3. The product team can control costs. Why can't it control the selling price?
4. What is the executive's role in product development?
5. How can project management help product development?
6. Recall the development of a new air emissions monitor presented in this chapter. Why would a purely Agile approach *not* work for this product's development? Are there instances where Agile is not the best approach?
7. There are fundamentally four phases of project management. Can you name them?
8. What is the difference between a business case and a project plan?
9. When is product development completed?
10. How important is the product/project manager's leadership skills?

Discussion Case 5.1 Questions and answers

1. You are brought in as a consultant to turn around this failing new product development program. What would be some of your first actions?
   This will be a long-term engagement and will not be solved or turned around in a month or even several months. Remember, a program is a group of related projects. There are 12 individual projects in this program. First actions you might take include:

Put all projects on hold for 30 days, until an assessment can be performed.

- *Meet with CFO to determine where program is off track. Which areas are over-spending? Which projects? are over budget*
- *Meet with each project team leader, individually, to assess where they are in their project plan and schedule. Obtain their thoughts on why they think their project is floundering.*
- *Meet with leaders of engineering consulting firms (individually) to obtain their thoughts and ideas to solve the problems. Many times, outside firms have a good idea of why things are not working.*
- *Meet with the head of R&D. What is holding them back? What type of assistance do they need to get back on track?*
- *Meet with the leader of construction group to obtain their input.*
- *Determine if scope changes are contributing to problems and why they are occurring.*
- *Discussions with individual team members should allow a picture of issues and problems to become clearer. It is also likely that problems in the various projects are impacting or exacerbating other projects. This will enable you, the consultant, to call a full team meeting to discuss findings and discuss development of new program and project plans. Because the program is in the ditch, it is likely that poor planning is a root cause.*

2. The three project leaders (director-level employees) are firmly entrenched in this program. They will most likely resent your presence. How would you deal with this?

   It is important that you and your team exhibit integrity and have a heart-to-heart discussion with each director. It comes down to trust: how do they know they can trust you? Fundamentally, the director's will be concerned that they will look bad if an external resource can solve their problems for them. After all, why would their company need them if they cannot solve the problems they are being employed to address?
   Here are several techniques that can get them to become allies instead of enemies:
   Indicate that you are there to help them succeed.

   - *Give them the tools and share your expertise. Be a mentor and coach.*
   - *Give them a lot of the credit when success arrives.*
   - *Gain their friendship: join the company softball team, take key people to lunch once a week, play golf with them, etc.*
   - *Take an interest in them as people, not just the job description.*
   - *Be empathetic. This requires a leader with a high emotional Quotient (EQ).*

3. Would you bring in any outside talent to assist you? If so, who or what kind?
   A program of this size and complexity will require a consulting team. It will be very important that your team has qualified and credible people to work with and mentor the various project leaders. A project this size would need from three to five consultants.

4. Once you have a clear picture of the program's true status, what would be your next steps?
   Since the project is on hold for 30 days, this provides time for each project team to develop a new and detailed project plan. However, each team should follow the same "Table of Contents" (presented earlier in this chapter) so that they are all consistent.

Each project plan should be signed-off by the team leader, program manager and product manager (if different).

A new plan will refocus the teams and reset expectations with the executive team. The new plan will layout a new budget and schedule.

An overall program plan should also be developed. This describes how the individual projects impact each other, how they will be coordinated together, and the overall budget and schedule (combination of all 12 projects).

5. How would you propose to get back on track?

   - *Once new project and program plans have been developed, track these to ensure that all activities are occurring on time and budget.*
   - *Status reporting should occur monthly.*
   - *Regular team meetings to determine roadblocks and how to remove them.*
   - *Perform risk reviews on a regular basis.*
   - *Set up a "war room" as a dedicated center of control.*
   - *Ensure all project leaders understand the accountabilities and project objectives.*
   - *Ensure that all project participants know their roles and responsibilities.*

6. What are your chances of success?

   Assuming the market demand for the products is still strong, there is a very good chance of success. The program is still in the early planning stages. Success will also be determined by adhering to the "new" budgets and schedules and management's understanding that expectations are to be reset.

7. What are some techniques you could use to improve communications?

   - *Regular status reports*
   - *Individual one-on-one meetings with team leaders as needed*
   - *Weekly progress meetings (include suppliers and consultants)*
   - *One-on-One communications weekly with executive sponsor.*
   - *Have an open-door policy so team members feel comfortable in bringing problems to your office*

8. How will you measure success and/or progress?

   - *Monitor the budget and schedule for variances.*
   - *Have regular "gate" meetings where the PM must justify their progress and/or actions.*
   - *Be cognizant of major milestone dates. Are we meeting these?*
   - *Maintain vigilant contact with suppliers to confirm shipping dates.*

9. Trust is very important in any consulting engagement. What ideas can you implement to promote yourself as trustworthy?

   - *Establish relationships with key client people. This will take time.*
   - *Be interested in your client's concerns.*
   - *Be proactive in providing solutions to problems.*
   - *Coach and mentor your client's personnel.*

10. While there are many project management (or lack of such) issues related to this program, there are also many organizational problems. What can you do to help the organization?

    The organization has many issues which need to be addressed. As a consultant, you can recommend changes, but actual change must come from the top of the client company – generally the executive suite. As an outsider, you do not have the authority to make these types of changes. Here are a few things to recommend to the client's CEO:

- *A program must have one person in charge. Having a critical program run by three people is not practical.*
- *People must be held accountable. Those who are not performing should be replaced.*
- *The program is split into three locations. Both Headquarters and Administration should have a representative located full-time at the manufacturing site. This will improve communications and supply chain functions between the three locations.*
- *The person appointed to lead the effort should have excellent managerial and leadership skills. The technical issues can be addressed by the R&D or engineering teams. It is more important to have managerial skillsor this particular role.*

Answers to Discussion Questions

1. Why is the WBS important?
   The WBS represents the total scope (or breadth) of the project.

   - *For each task identified, a cost can be determined, and each task cost can be summed to provide a total project cost.*
   - *For each task, a person or group can be assigned responsibility.*
   - *A project schedule can be developed by importing each task to the schedule.*
2. Agile has become very popular in the software development industry. Will this displace traditional project management? Provide reasons to support your answer.
   Because Agile is an iterative approach, it is valid for developing software systems or technology applications. It involves customer and supplier input and support and develops a minimal initial product at each stage of development.

   Agile would not be very effective for long-term capital projects such as facilities improvements, developing "hard" products such as cars, infrastructure, or manufactured goods. Some items cannot use an iterative approach as each change can amount to large cost or schedule impacts (negative). Essentially, traditional PM discourages changes beyond a certain point while Agile encourages changes throughout the project lifecycle.
3. The product team can control costs. Why can't it control the selling price?
   The product or marketing team can propose a selling price, based on research and customer input, and the required margin for profitability, but it will be the market that ultimately determines the price. If your firm releases a new product, and competitors can quickly compete, your price will be lower. If, on the other hand, your product is truly unique, with few competitors, you may be able to charge an even higher price.
4. What is the executive's role in product development?
   A key senior executive should always be appointed to support a new product or project. He or she should provide funding, resolve conflict at a higher level than the product team, serve as a guiding resource, and be a link between the product team and senior management.
5. How can project management help product development?
   The principles of project management are fundamental to project success. It will force the team to do detailed project planning, monitor and control budgets and schedules, communicate more effectively, evaluate potential problems (risks) before they develop and provide a structured approach to both planning and execution.

6. Recall the development of a new air emissions monitor presented in this chapter. Why would a purely Agile approach *not* work for this product's development? Are there instances where Agile is not the best approach?

   The air emissions monitor is a "hard" product manufactured in a production environment (refer to Question 2). The principles of embracing change are not beneficial for a manufactured product, which resists change.

7. There are fundamentally four phases of project management. Can you name them?

   Initiating, Planning, Execution, and Closing. PMI° would add a fifth element, monitoring and control across all four phases. This means monitoring and controlling budgets, schedules, risks, etc. from beginning to end. The business case is not considered one of the fundamental phases; it should be completed before a project is approved and funded, i.e., before project initiating.

8. What is the difference between a business case and a project plan?

   The business case is an initial feasibility study. It is used to provide information so that the executive team can make a go/no-go decision. The business case also evaluates several alternatives with recommendations provided. The business case may also offer high-level or preliminary budgets, timelines, resources, etc. However, a detailed a business case cannot be at the depth or appropriate level of action compared to a project plan. A project plan, on the other hand, is a document that describes how the new product will be developed with detailed planning, logistics, execution, etc. The project plan takes the business case to a much higher level of detail.

9. When is product development completed?

   Generally, when the product is in regular production, that meets QA/QC standards, and shipped to customers.

10. How important is the product/project manager's leadership skills?

    The product manager/project manager must have a number of leadership skills to be successful. These include communication skills, negotiating, instilling trust, resolving conflict, ability to make decisions, and motivating the team. These are all hallmarks of leadership.

Notes

1 https://angelsanddemons.web.cern.ch/faq/what-does-cern-mean.html.
2 https://home.cern/about/who-we-are/our-mission.
3 https://www.britannica.com/event/Human-Genome-Project.
4 A guide to the Project Management Body of Knowledge, the Project Management Institute, 6th ed., Newtown Square, PA, 2017.
5 Mark C. Layton, Steven J. Ostermiller, Dean J. Kynaston, Agile Project Management, (2020), John Wiley & Sons.

Bibliography

Katzenbach, J. and Smith, D. (1993). *The Wisdom of Teams*. New York, NY: HarperCollins.

Kynaston, D., Layton, M., and Ostermiller, S. (2020). *Agile Project Management*, 3rd e. Hoboken, NJ: John Wiley & Sons.

Lewis, J. (1995). *Project Planning, Scheduling & Control*. McGraw-Hill.

Mantel, S. and Meredith, J. (2000). *Project Management – A Managerial Approach*, 4th e. New York, NY: John Wiley & Sons.

Project Management Institute (2017). *Agile Practice Guide*. Newtown Square, PA: Project Management Institute, pub.

Eric Ries, The Lean Startup, Penguin Random House, (2011), New York, NY.

6

Product Development for Small Firms and Entrepreneurs

Much of this book has been directed towards large companies with significant financial resources and staffs of talented people. But, what if your firm has only 100 people and limited cash resources? Or only 50 people? Or a startup with six people? What would be a practical approach for product development? How can one compete with the big corporations with billions in annual revenue? Where can small companies go for help?

Funding for Your Start-Up: A Necessary Ingredient

Today's small companies and start-ups have many options to consider in how they develop their products. However, it should be noted that most startup companies fail not to a lack of ideas or creativity, but to a lack of capital ... that is, they run out of money. There have been several successful firms where the founders had extremely high drive, believed in their product, and were able to sell their ideas to a market or partners. Microsoft and Apple were both started in their founders' home garage and today they are earning billions of dollars in revenue annually. Of course, this did not happen in one year.

But you must be realistic in your quest to develop your product: if will take cash to get started and succeed.

Start-ups basically have five options to obtain funding for their products and ideas:

1. Use their own personal savings; and convince friends and relatives to provide additional funding as passive investors.
2. Obtain a loan from a bank or the Small Business Administration (SBA).
3. Seek funding from venture capital groups.
4. Go public and issue shares of stock.
5. Reach out to Angel Investors.

The first one, using your personal and family savings, is the easiest to obtain. The remaining four have obstacles. Although, it may not be enough.

Product Development: An Engineer's Guide to Business Considerations, Real-World Product Testing, and Launch, First Edition. David V. Tennant.
© 2022 John Wiley & Sons, Inc. Published 2022 by John Wiley & Sons, Inc.

Loans from the Bank and Small Business Administration (SBA)

Banks have an aversion to lending money to those with no assets to put up as collateral. What are assets? This means you own a building, capital equipment, or even your home, that can be put up as collateral. Should your business fail, these items become the property of the bank to dispose of as needed to get their money back.

On the other hand, say you wish to borrow $100,000 but as a show of good faith you will also invest 25% of that amount, $25,000, of your own money – investing a total of $125,000. You now have a stake in your company succeeding. The bank may be impressed with your sincerity and commitment, but they will want to see a detailed business plan. What is the product to be offered? What is the market? How do you know this market exists? They will also want to know the track record and credentials of the company founders. Have you run a successful business before? What is your background and educational level? Do you understand business financial documents such as a Statement of Cash Flows or Income Statement, etc.?

Without a background in business or a track record in start-ups, it will be difficult to obtain financing from a bank. Not impossible, but difficult. Even if the bank decides to provide some funding, and even with some assurance of success, you may find your interest rate higher. This is how the bank will reduce its risk. An alternative to a loan is a line of credit. This will still have an interest rate but will provide access to capital on an as-needed basis.

The author is aware of entrepreneurs that put up their homes as collateral to obtain funding. Some were ultimately successful; others were not.

However, it is not all negative. With enough preparation, a solid business plan, and a great management team, there are banks that work with small businesses. And, at the time of this writing, interest rates are extremely low.

The Small Business Administration (SBA)

SBA loans are partially guaranteed by the Federal Government and are administered through various financial institutions (banks). The SBA offers competitive rates, support in the way of education and advice, and offers different types of loans based on your business. Like other lenders or partners, there are qualifications that a business must meet for loan eligibility. Here are several notes regarding of SBA-backed loans:

- The business is legally registered.
- It operates in the USA or its territories.
- The business owner has invested time and funds in the business.
- The business is *unable* to obtain funds from other lenders.
- The SBA lowers risks and provides easier access to capital.

Information about SBA can be found at:
https://www.sba.gov/funding-programs/loans

Funding from Venture Capitalists

First, what is a Venture Capitalist (VC)? Venture capitalist firms are in the business to make huge ROIs by investing in small companies or startups that are ready to commercialize their product, service, or idea. They generally do not begin "out the gate" with startups; rather, they wish to invest in a company and provide equity (cash) in exchange for part ownership. Therefore, if you wish to obtain funding from a VC firm, you will lose some control of your company. This will depend on the terms that you negotiate and agreement reached with the VC (Ownership split 60/40? 50/50?). Typically, VCs will provide funding, but will want to cash out with a huge ROI (Return on Investment) at some point in the future. This could be a timeline of five or ten years. Both Facebook and Twitter, for example, were funded by VC money. It would be prudent for you to have an exit strategy. This means you are prepared to sell your business at some point in the future. This is not a bad situation, especially if your firm has become extremely profitable. Many entrepreneurs walk away with significant funds and then go start another business.

The VC firm will provide advice and nurturing to have a greater chance of success. Because VCs invest in smaller companies, the failure rate can be quite high, but one big payoff can obliterate any losses incurred from other, less profitable firms.

A VC firm is generally incorporated as a Limited Partnership (LP) where the partners provide funding. A VC committee meets to decide in which companies they will invest. The VC firm is paid for their work in the form of management fees, but the majority of profits will flow back to the partners; that is, investors.

Where Do VCs Get Their Funding? Who are Their Partners?

A VC firm is funded by limited partners such as large pension groups (teachers' pension funds, for example), insurance companies, wealthy individuals, corporate pensions, and foundations. Their investment cash is pooled together, and the VC firm, as the general partner, manages the funds and investment decisions.

Before investing, the VC firm will seek companies with a strong management team, a unique product that will have a huge ROI potential and few competitors. This is known as having a large moat. The moat, in medieval times, protected the castle from enemies. In the business sense, it means having a product that will take competitors years to duplicate and who may not have the efficiency or knowledge to compete effectively. A large business moat means you have advantages that protect your investment, product lifespan, and profitability. As stated previously, VCs are looking for firms that are ready to commercialize an idea or product. If your firm has not put together an effective business case, done significant research and design, performed market research, nor built a prototype, it will be hard to get a VC firm interested in your company.

The bottom line is VCs provide funding for potential (i.e., risky) projects and are an alternative to the bank. But your company will need to already have some success or be further along with a great idea or product. Remember also, that a VC will want to flip the business at some point. They wish to obtain their profits by "flipping" your company, that is sell it, and invest the funds elsewhere in other startup firms with potential.

Funding by Issuing Shares of Stock

Issuing stock for public sale – an Initial Public Offering (IPO) – carries its own risks and benefits. Like VC firms, you are giving up some ownership of your company. Most entrepreneurs will want to maintain a majority stake: 51%. Even if you decide to maintain 51% ownership as the primary shareholder, the other 49%, shareholders, own almost one half of your company and have certain rights. They have the right to vote for (or against) the board of directors. At some point, they will expect your company to pay dividends. Further, your company, the board of directors and executive management, will be required to abide by federal laws (Securities and Exchange Commission – SEC), and other requirements. Sarbanes-Oxley is one of those laws. It was passed in Congress in 2002 due to several highly visible firms rigging their financial statements. This duped investors, both big funds as well as individual investors when the firms went bankrupt. The first BIG firm to do so was Enron, an energy services company that, at one time, was worth $63 billion. Several other large publicly held firms, among them WorldCom ($103 billion), and Tyco.

As you might expect, becoming a publicly held company is no small matter. Therefore, small firms generally do not have the legal or business expertise to become listed on the stock exchanges (NYSE or NASDAQ); and, if your company's stock price slips below $1 per share, your firm will be delisted. Going public with your firm is not for the faint of heart, but once you have established your product, a VC firm can assist in taking your firm public as can any good business attorney. It is customary for entrepreneurs that take VC money to have an exit strategy. After all, the VC will want to flip the firm at some point.

Funding with Angel Investors

Angel Investors are different from VCs as the partners invest their own personal wealth. These high net-worth individuals are interested in seeing startup companies succeed and do not forget, they also wish to make money for which they are passionate. Some investors like to champion startups that will promote social good or solve some of the world's problems. This can include the environment, healthcare, education, etc.

Because many angel investors have been at senior levels in large companies, they can offer experience, advice, and facilitate key business introductions to startup founders. As with VCs, angel investors look for an exceptional management team with a great product. Further, a startup that becomes successful will provide hundreds or perhaps thousands of jobs which helps fuel economic growth. Figure 6.1, Angel Investments in 2019, shows the most common types of investments made by Angel Investors during that year.

It should be noted that Angel Investors have become a large source of funding for startups – one source indicates that over 67,000 companies annually are funded by Angels.[1]

How does one find an Angel Investment group in your location? According to Hockeystick/Angel Capital Association,[2] there are approximately 14,000 Angel

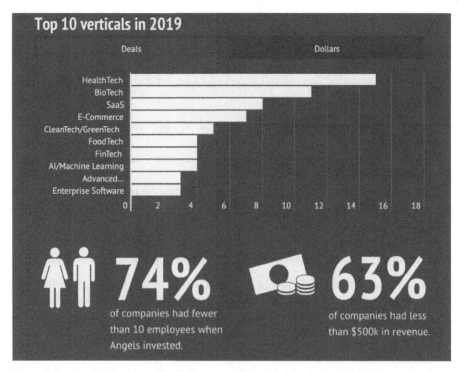

Figure 6.1 Angel Investments 2019. *Source:* Hockeystick/Angel Capital Association Aug 2021

investors belonging to 275 Angel groups. The Angel Capital Association has a nation-wide (US) directory listed on their website. There is a high probability that several are near your location.

https://www.angelcapitalassociation.org/directory

Other Sources of Business and Financial Assistance

All startups should seek and apply to join a business or technology incubator. What is an incubator? For starters, there are over 500 incubators spread across the United States. An incubator is an organization that exist to offer advice, assistance, coaching, mentoring, expertise, education, and business connections. Many are associated with major universities or funded by state economic development programs.

Draper University, from their website,[3] offers this definition of their program:

> Business incubators are specially designed programs to help young startups innovate and grow. They usually provide workspaces, mentorship, education, and access to investors for startups or sole entrepreneurs. These resources allow companies and ideas to take shape while operating at a lower cost during the early stages of business incubation. Incubators require an application process to join and usually require a commitment for a specific amount of time.

All incubators are committed to helping viable startups succeed. Some, as is the case with the Advanced Technology Development Center at the Georgia Institute of Technology (Atlanta), offers education and coaching, access to the technical resources within the university, and exposure to investor groups.

In general, incubators provide an exceptional service for startups. Below are a few elements of which to be aware:

- Some incubators specialize in a specific industry, such as technology, medicine, manufacturing, etc.
- Incubators can provide networking to other companies, successful entrepreneurs, and investors.
- Some are free of charge, while others charge a nominal fee to ensure you are serious about your company.
- Most will require a time commitment.
- Not everyone who applies to an incubator program is accepted.
- The company founder(s) must demonstrate a passion for their product or idea.

Joining an incubator will give you an edge and will force you to focus on your company's purpose, product, and logistics. Many times, entrepreneurs have a general idea of what they'd like to develop, but the incubator will help you drill down and prepare you for meetings with investor groups. Below is a listing of incubator groups by state.

Incubators by State
https://www.gaebler.com/Business-Incubator-Lists-By-State.html
A few selected incubators:
University of Texas, Austin
https://ati.utexas.edu
Georgia Institute of Technology, Advanced Technology Development Center, Atlanta
https://atdc.org
University of Chicago
https://polsky.uchicago.edu/programs-events/new-venture-challenge
Angelpad, New York City
https://angelpad.com
500 Startups – Silicon Valley (San Francisco, CA)
https://500.co
California State University, Apostle Incubator, Long Beach, CA
https://www.csulb.edu/institute-for-innovation-entrepreneurship/apostle-incubator

Summary on Product Development and Sources of Funding

It may sound very positive that lots of money is available, just waiting for entrepreneurs and startups to walk in the door. However, this is not as easy as it may sound. Any investment group, regardless of type, has as its primary mission to make a huge ROI.

Consequently, a startup needs to have the following consideration in place for any investors to consider your firm for funding:

- A strong, experienced management team
- A well thought-out business plan

- A clear, concise reason for your firm to exist
- A SWOT analysis (usually contained in the business plan, see also Chapter 2)
- A review and evaluation of competitors (if any)
- A financial model evaluating the potential costs and profitability of your product (including the break-even point if a product or service)
- Your firm is advanced far enough that you are ready to ramp up commercialization
- An excellent sales pitch and value proposition
- Your product has a strong moat
- Your product has been tested and works.

Clearly, this is an ideal "wish list" for investor groups. It is unlikely that firms will have all of these criteria in place, but it is a starting point.

Even if interested, the potential investment group will cross-examine you with tough questions, attempt to poke holes in your business plan, and grill you on why they should invest in your company versus the many other equally worthwhile startups that are out there.

You should expect a tough grilling, but if you prepare and you have a great product or idea, then you will have as much opportunity as other startups. However, you can greatly increase your chances by working with an incubator as previously discussed. It is also worth noting that most investment groups want to see your product beyond the startup phase: that is, expansion to start making significant dollars with enough capital investment. In contrast, Angel Investors are willing to be more flexible with financing terms or percentage ownership and, *in some instances*, are willing to fund startups at the beginning stages.

Small Firm Challenges

Small companies have more challenges and obstacles to confront than big firms. For one, their budgets are smaller, in-house expertise may be limited, funding is constrained, and there is less clout in the business community. Further, some firms may not have the relationships – this is especially true in the global economy. Some of the challenges faced by small firms are barriers to entry, others are just inexperience or not having the required competencies. Some of these in the below list have been previously mentioned:

- Lacking internal talent for engineering, design, and testing
- Do not understand how to develop business plan or cost-benefit models
- Lack of capital
- Lack of structured planning (project management skills)
- Inability to execute on ideas
- Marketing message not strong or clear
- Legal and regulatory obstacles
- Do not understand the market or competitors.

While the above "obstacles" may seem daunting, as a small firm there are some advantages that you can employ. Big corporations face the same challenges, only they have more bureaucracy. And, yes, big corporations have lots of resources (people, equipment, money), but as a small firm, you have the advantage of nimbleness and zero bureaucracy. Let's look at each obstacle above (other than those already discussed) and determine how to overcome them.

Lacking Internal Talent for Engineering, Design, and Testing

First, it would make sense to perform an internal "competency" review, perhaps even a SWOT review. What strengths reside within your company? What are your core competencies? What are your firm's weaknesses? Once you can answer these questions, you can take steps to address them.

If you do not have the technical talent in-house, there are many consulting engineering firms and further, independent manufacturing companies that also employ engineering talent. It may be that you should outsource the engineering and testing to one of these companies. Also, there are many freelance engineers that can provide their expertise in the technical aspects of developing your product. Freelance engineers will be less costly than a full-service engineering firm.

How do you find freelance engineers? There are many technical societies, some of them previously listed: American Society of Mechanical Engineers (ASME), American Institute of Chemical Engineers (AICHE), etc. Each of these societies have local chapters and monthly meetings. Joining one of these or attending their events is one way to meet engineers. All of these societies have annual conferences that include trade shows. Trade shows are a rich source to find talent or even suppliers you have not considered.

Do Not Understand How to Develop Business Plans or Cost-Benefit Models

Working with an Incubator would provide your firm with access to training, experienced entrepreneurs, and other contacts. As previously discussed, Incubators are an exceptional source for practical advice and business contacts. They can assist you in developing a business plan; also, many MBA students may be able to assist at a very reasonable cost. It may be worth contacting a local university that offers an MBA program and offers this type of service.

Lack of Structured Planning

As discussed elsewhere in this book, project management skills are strongly advised for any company, but especially startups. Key topics that are presented in this book include:

- Risk reviews
- Budgets and schedules
- Communications and coordination
- Quality Assurance and Quality Control
- Managing resources (people, money, equipment)
- Defining the scope or breadth of your initial product
- Teamwork and Leadership

The reader should review the pertinent sections on these topics.

Inability to Execute on Ideas

Structured and detailed planning will allow your firm to execute fully and efficiently. The purpose of the structured plan is to not only outline the scope, budget, and schedule, but to implement them while monitoring your progress. It serves as your project roadmap.

For instance, the schedule should contain all the activities or tasks with start/end dates needed to bring your product to fruition. Milestones should also be a part of your schedule. If you are slipping key milestone dates, then this indicates you are falling behind, and that corrective action is needed to get back on track. The schedule also keeps your attention focused on activities and their progress.

Marketing Message Not Strong or Clear

As a startup, you may not have a marketing department. Perhaps your product is not ready for promotion. And, traditional marketing approaches, while still valid, have been somewhat displaced by visibility through online and social media platforms.

At some point, you will need to promote your product and marketing will become a key part of your efforts. Clearly, there are marketing firms that can provide marketing services. Yet, like engineers, there are many independent and freelance marketers that can assist you.

Legal and Regulatory Obstacles

There are always standards and possibly regulatory requirements for certain products. In the USA there are standards relative to engineering materials, pressure products (codes), and electrical safety (grounding for common electrical appliances or products such as hairdryers, computers, refrigerators, etc.). Experienced engineers will be aware or familiar with them. Regulatory requirements are usually federal or state laws with which one is required to comply. For example, our air emissions monitor would need to measure pollutants within federal pollution guidelines. Failure to do so would invalidate the purpose of the monitor.

If your product will be sold offshore, there will be a different set of standards. For example, most of Europe's household electrical supply is 240 volts at 50 Hz., as opposed to the USA's which is 120 volts at 60 Hz. Some Asian countries are also 240 volts. Your product may need to function on a different electrical standard offshore.

Another European standard is the International Organization for Standards (ISO). It publishes standards for Industrial, technical, and commercial organizations. As an international standard setting organization, you may find your potential product must comply. Many US companies doing business in Europe and other places (the standard is an "international" standard), comply with ISO standards. ISO is a grouping of standards organizations from several counties and is headquartered in Geneva, Switzerland.

Use of a Product Roadmap

Products evolve over time. This may be due to multiple factors including advances in technology, competitor pressure, changes in market or customer segments, changing consumer tastes, and product functional and technical obsolescence. It would be unrealistic to think the product you launched 10 years ago is still relevant and it may not be supported by the manufacturer today.

A product roadmap is like a product/project schedule, but it is based on future product revisions or iterations. Prime examples of future product changes include:

Car companies: Cars evolve over time due to several of the factors previously noted.

- Technology – car technology appears to increase exponentially with time. This includes driver-interface (touchscreens) and electronic modules to control engine dynamics and efficiency. Electric vehicles are a new departure and advancement from the traditional IC (internal combustion) engine, with a completely different and ground-breaking technology. EVs may be a "disruption" technology.
- Competition – car companies compete against one another by introducing not only technology, but nice "would like to have" features such as full-length sunroofs, new and appealing dash materials, enhanced displays, etc.
- Changing customer tastes – customers can get bored with the same look year after year. A car (or truck) is fundamentally a platform containing seats on four wheels. But each car maker has their own styling designs and market segments they appeal to. Every three to four years, car companies relaunch their car lines with new styling. Even if the engine, driveline, and suspension are the same, the body style must undergo refinement to keep the vehicles looking fresh and new.
- Market segments include luxury cars, sports cars, sport utility vehicles (SUVs), rugged off-road vehicles, family cars and vans, pickup trucks, and others.

It must be noted that regulatory changes will also drive the adoption of technology. For example, the car industry, besides competitive issues, has been forced to comply with new regulations related to air quality and environmental climate change. Federal policy in the United States has been penalizing IC engine cars through higher gas taxes while offering generous tax incentives for those purchasing electric vehicles.

Some countries have mandated that IC engines in new cars are forbidden after the year 2030. It should be apparent that governmental policy can shift or stimulate new product development.

Other competitive industries would include smart phone manufacturers, smart TVs, large home appliances, apps for phones or computers, computer manufacturers, etc. In essence, every producer of products must continually update their products or services to stay in business. If your company is small, you will still need to update your products and services over time; otherwise, your competitors (big and small) will pass you by. Have you thought about future revisions for your product?

So, how does a product roadmap relate? First, what is a product roadmap? A product roadmap outlines the strategy and planned evolution of a product. It allows the product manager to work with the development team as to which features will be incorporated into the current product over time.

A product roadmap can be useful in presentations to executive management, or potential and existing customers. Documenting a planned strategy and the activities to deliver new features over time is useful. This approach should help you obtain senior management consensus, continued funding, and possibly obtain advance orders from customers.

Below are items to consider as your roadmap is developed:

- What are your competitors doing? Can you leapfrog around them or stay in front? Will the revised product have the same target audience?
- What features and enhancements do our customers want? How will we determine this?
- What timeframe (i.e., how far out) do we wish to go?
- Will the same design team be available?
- Internally, how will we proceed? How will new ideas and features be communicated to team?
- When should draft and final plans be submitted for executive signoff?
- At what point is the revised product revealed to the public?

Figure 6.2 Product Roadmap – Our Air Emissions Monitor, shows a high-level example of a product roadmap that will be revised approximately every three years. This type of monitor will see significant technology advances over time and therefore requires the manufacturer to keep abreast of these advances and implications for their product. This would be a combination of both Marketing and Engineering professionals.

For highly technical products, such as the monitor, it would be beneficial for the company to keep the same team engaged for all future refinements. Not only will the team be familiar with the product, but also how it was developed, current market conditions, technology leaps, and customer preferences. It is a competitive advantage to keep people familiar with the product on the team rather than put together a new team every three or four years.

Figure 6.2 is a brief timeline of an air emissions monitor's evolving development. The team can be as detailed or as brief as needed. The timeline was derived using an excel spreadsheet. For 15 or 20 activities, this is adequate. If the team went into more detail and had, for example, 50 or 100 activities, it would make more sense to place the activities into a formal scheduling program such as Microsoft Project™.

Innovation

Suggestions for Success

It must be emphasized that the following factors are necessary to be successful:

- You new product must be unique and interesting. Who wants to see another "Brand X" widget that is already offered by many competitors?
- Customer input (desired features) must be considered in new product development or refinements. Take the time to speak with potential and existing customers and value their feedback.
- The customer must receive (or perceive) value for the money spent.

Future Activities by Group — Product Roadmap

Future Activities by Group	Year 1 Q1	Q2	Q3	Q4	Year 2 Q1	Q2	Q3	Q4	Year 3 Q1	Q2	Q3	Q4
Marketing												
Gather customer wants/needs	■	■			P				■			P
Develop priority list for next release:					R	■				■		R
Styling			■	■	O					■	■	O
Technology Advances			■	■	D	■	■			■	■	D
New display (Coord w/ Engineering)			■	■	U		■			■	■	U
New customer interface - customer preferences				■	C		■			■	■	C
New compact size				■	T		■			■	■	T
Work w/ Engr. re customer requirements												
Engineering												
Engineering reviews customer wants/needs	■				R	■				■	■	R
Brainstorming and team discussions about ideas					E	■	■			■	■	E
Engineering creates preliminary designs	■				L		■			■	■	L
Reviews results with Marketing and team	■				E		■			■	■	E
Risks review evaluation					A						■	A
Cost-production analysis					S						■	S
Meet with production to discuss Manufacturing					E						■	E
Production												
Meets with Engineering to review product updates			■		P		■			■	■	P
Revise production flow drawings			■		R		■			■	■	R
Determine machine adequacy for new product			■		O		■			■	■	O
Meet w/ Eng. and Mktg. to discuss costs			■	■	D			■	■	■	■	D
Determine needed material inventory levels				■	U			■			■	U
Confirm production schedule					C						■	C
					T							T
Advertising												
Prepare PR materials - discuss & Review				■	R			■	■	■		R
Confirm collateral material is accurate				■	E				■	■		E
Print and distribute					L				■			L
Update social medial platforms					E				■		■	E
Update company website					A				■		■	A
Sales												
Meetings to discuss customer needs/wants		■			S	■				■		S
Feedback on prototypes and ease-of-use			■		E		■					E
Develop story for customer applications			■				■					

Figure 6.2 Product Roadmap. Figure by David Tennant

- Speed to market is important: first to market usually gets market share.
- Use project management techniques to keep your new product's development on track.
- You cannot do it alone. Seek partnerships, learn from others' mistakes, join an incubator, and seek guidance wherever you can. Networking with business leaders, finance groups, chambers of commerce, etc. is useful in making progress with your company. It is likely you will find contacts and expertise that are willing to help you.

When (Or If) to Patent

There are some who subscribe to the idea that patenting your process or product puts your invention "out there" in the public domain. Anyone can initiate a patent search and potentially find your new product. Would they copy it? Perhaps. But it would have to be different from what you submitted in your patent application.

Some of the questions an entrepreneur or inventor should ask:

- Do I need a patent attorney?
- Is my invention/product/process even patentable?
- Should I apply for a patent?
- What protection does an NDA (Non-disclosure Agreement) offer as opposed to a patent?
- What type of patent should I seek?
- When should I apply for a patent?
- How much will it cost?
- How long does it take to obtain a patent?
- How long is a patent good for (enforceable?)?
- My invention or technology is new and disruptive. Will the Patent Office be able to understand my patent?
- If I wish to patent my invention overseas, what is the process and costs?
- What are the benefits to having a patent?

These are all valid questions that any inventor should ask. These are addressed below.

Do I Need a Patent Attorney?

The answer is yes if you wish to make sure the application is done correctly. You may be able to file on your own, but history is littered with patents that were contested or lawsuits filed years after the original patent application. It is much more troublesome and expensive to fix a problem later.

Is My Invention/Product/Process Even Patentable?

A patent attorney will be able to evaluate your product and offer advice. Most patent attorneys also have degrees in science or engineering, so they can understand technical issues and processes.

Should I Apply for a Patent?

Before an application is drafted or submitted, an attorney can offer advice and the chances of your product obtaining a patent. Part of the decision will be based on how close to commercialization your invention is. Has a prototype(s) been built and tested? How far along in the design or production process is your product?

Based on preliminary discussions, an attorney will also provide guidance on the type of patent you should seek.

What Protection Does an NDA Offer as Opposed to a Patent?

Many times, an entrepreneur or small company will work with outside suppliers such as engineering firms, materials vendors, manufacturing companies, etc. to develop a product or process. An NDA is useful in that it prohibits suppliers and others from disclosing sensitive or proprietary information. However, it is always possible that key information may be leaked inadvertently or on purpose. Usually, an NDA will have a termination period – generally within two to three years after signing. So, an NDA will offer short-term protection.

A patent has more legal protection and enforceability, and of course, is good for a much longer timeframe.

What Type of Patent Should I Seek?[4]

There are several different types of patents, depending on the object or item you wish to patent. These types are defined by the US Congress. Below is a description of each

- **Utility patent.** This a detailed document with the intent to educate the public how to use a new machine, process, product, or system. New technologies such as genetic engineering, internet-delivered apps or software, are pushing the boundaries and definitions of what kinds of inventions are eligible for a utility patent. New materials or compositions of materials also fall under utility patents.
- **Provisional patent.** US law allows inventors to file a less formal document that proves the inventor was in possession of the invention. Once filed, the invention is "patent pending." This has a one-year deadline to file for a formal patent, otherwise, the provisional patent expires.

  It should be noted that listing "Patent Pending" on a product or advertising materials also provides a marketing and PR advantage. This promotes your product as "new" and possibly more advanced than your competitors. Many people like to purchase the "latest" products.
- **Design patent.** This patent provides protection for the appearance and design of an object or item. The shape of a bottle (think Coca-Cola) or the design of a shoe (e.g., Nike), can be protected by a design patent. In recent years, software companies have used design patents to protect elements of user interfaces and even the shape of touchscreen devices.

  In the distant past (2011), Apple sued Samsung over patent infringement of its iPhone and iPad products. Essentially, Apple claimed that Samsung's smart phones looked the same and had similar features which were originally developed by Apple (including the shape of the phone with "rounded" corners). A jury awarded Apple $1 billion in damages, which is less than the $2.5 billion Apple was seeking.

- **Plant patent.** A plant patent protects new kinds of plants produced by cuttings or other nonsexual means. Plant patents generally do not cover genetically modified organisms and focus more on conventional horticulture.

However, genetically modified plants or seeds are completely new, lab grown plants can receive patent protection from UPOV – the International Union for the Protection of New Varieties of Plants. UPOV is an intergovernmental organization with head-quarters in Geneva (Switzerland). Founded in 1961, A number of countries are signatories including the United States, Germany, France, Russia, China, the United Kingdom, and many others.

> UPOV's mission is to provide and promote an effective system of plant variety protection, with the aim of encouraging the development of new varieties of plants, for the benefit of society.[5]

https://www.upov.int/portal/index.html.en

When Should I Apply for a Patent?

If one applies too early for a patent, then later development of the product may require filing again. Many times, it is worthwhile to apply when the product is ready for commercialization. This may mean after a prototype has been successfully tested and full production is to begin. It could also mean, in the case of software, that it has been thoroughly beta tested (client tested) and is ready for download sale to the public.

https://www.uspto.gov

How Much Will It Cost?

The costs associated with a patent attorney will vary. First, while there are many large law offices that offer this service, there are also many smaller firms that do the same work and will be less expensive. However, regardless of the legal firm, one can expect the cost to be minimally several thousand dollars ($5,000 to $10,000 range). This will also depend on how complex the new product or process is to understand. If an attorney must contract with outside expertise to evaluate a highly technical product, the costs of course would be more.

There is also a yearly patent fee for European patents. In the US, patent fees are due on a schedule of 3.5 years, 7.5 years, and 11.5 years. Failure to pay the fees will cause the patent to expire.

How Long Does It Take to Obtain a Patent?

A quick process from application submittal to granting a patent would be 18 months. Most applications take two to three years. Note, that it may take several years to develop your product, for example, a new vaccine. The patent would be applied for some time after successful trials – adding another three years.

How Long Is a Patent Good for (Enforceable)?

For utility patents (product, machine, process), the length of time is 20 years. For design patents, the timeline is 15 years.

My Invention or Technology Is New and Disruptive. Will the Patent Office Be Able to Understand My Patent?

The patent office has many resources with which to evaluate patent application products. The USPTO has several technology centers with competency to evaluate highly technical products or systems.

If I Wish to Patent My Invention Overseas, What Is the Process and What Are the Costs?

If you wish to patent your invention internationally, be prepared for large costs. The USPTO (United States Patent and Trademark Office) issues patents that are only recognized in the United States and its territories. If you wish to market your product or invention overseas, you will need to apply for a patent in each country. Obviously, it would be advised to use a patent attorney in each country that is familiar with the legal issues and patent processes. With 195 countries, the bureaucratic hurdles and costs could become excessive very quickly.

One group of countries that can assist is the EU (European Union). Like the US, where you do not need a patent in each state, the EU has a single patent process for the 27 countries that comprise the European Union. The EU Patent Office accepts applications under the European Patent Convention (EPC) and the Patent Cooperation Treaty (PCT).

Similar to the US, there are yearly fees to keep the patent in place. Failure to pay the renewal fees will remove the patent.

What are the Benefits to Having a Patent?[6]

Advantages

- A patent allows you the right to stop others from copying, manufacturing, selling or importing your invention without your permission.
- You get protection for a pre-determined period, allowing you to keep competitors at bay. A US utility patent is good for 20 years.
- Alternatively, you can license your patent for others to use it or you can sell it. This can provide an important source of revenue for your business. Indeed, some businesses exist solely to collect the royalties from a patent they have licensed.
- Having a patent may give you more credibility and advantage when seeking investors.

Disadvantages

- Your patent application means the information about your invention is publicly available. It might be that keeping your invention secret may keep competitors at bay more effectively.
- Applying for a patent can be a very time-consuming and lengthy process. Markets may change or technology may overtake your invention by the time you get a patent.
- Cost – it will cost you money whether you are successful or not – the application, searches for existing patents and a patent attorney's fees can all contribute to a reasonable outlay. Note that not all patents have financial value.
- You'll need to remember to pay your annual fee or your patent will lapse.
- You'll need to be prepared to defend your patent. Taking action against an infringer can be very expensive. On the other hand, a patent can act as a deterrent

Chapter Key Points

- Small companies or sole-proprietor entrepreneurs' biggest obstacle to stay in business is cash flow. Companies simply run out of money.
- There are many options to finance a small business with a new product or idea. These include Angel investors, banks, venture capital groups, and issuing stock (becoming a publicly held company).
- Issuing stock or obtaining venture capital will take away some of the ownership of the company from the founders and to shareholders (stock) or partners in a venture capital firm.
- A good source of advice, education, and introduction to venture capital is through an incubator. These are generally funded through state economic development agencies or universities.
- Firms wishing to sell their products in other countries need to be aware of standards, quality guidelines, and regulatory requirements.
- A project roadmap is a timeline that will indicate enhancements or changes to your product over the next few years. It is meant to communicate to the team what changes will occur and when.
- As a product is developed, it should be released with future thinking. This can occur with a Product Roadmap: What features should be included in the next release? How often will a new revision or update be released? What is driving our updates? (Technology advances? Consumer tastes?).
- It is generally considered advantageous to have the same product development team in place throughout a product's development and subsequent updates.
- A patent may keep competitors from stealing or copying your invention. Be prepared to pay for a patent search, attorney's fees, patent application fee and recurring fees to keep your patent current.

Figure 6.3 outlines the patent process in the United States. The seeking of a patent can be a complicated process. Figure 6.3 shows a basic approach to obtaining a patent.

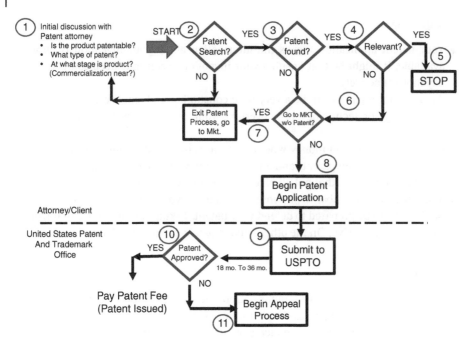

Figure 6.3 US Patent Process. Figure by David Tennant

Discussion Questions

1. At what point should a small firm or entrepreneur approach external funding such as venture capital or angel investors?
2. Some companies like to issue stock to raise funding. What are some of the upsides and downsides to this approach? At what point should a firm seek to become "public" and offer stock as an IPO?
3. You will be launching your new product in six months. How will your company and its product get visibility prior to launch?
4. Whether your firm is small or large, many product development teams are scattered geographically. For example, headquarters may be in Chicago (engineering and marketing), manufacturing facilities in South Carolina, software developers in California, and a second manufacturing plant in Indonesia. This makes teamwork, coordination, and communications more difficult. What are some of your ideas to overcome these obstacles? How will you build a team?
5. How will we know our product is ready for launch?
6. What can a small firm do if demand for product outstrips supply?
7. How would a patent attorney help your small company?
8. Who monitors the market for patent infringement?

Discussion Question – Answers

1. At what point should a small firm or entrepreneur approach external funding such as venture capital or angel investors?

If you believe you truly have a disruptive technology or unique product, seek those investors (Angels) who are willing to work early with small startups. It may be worthwhile to approach business associates for an immediate partnership. Otherwise, seek mainstream investor groups (or loans) when there is a track record in place for you product: initial customer tests have been positive and you need funding to expand production to meet anticipated demand. Be sure that you have a solid business plan.

2. Some companies like to issue stock to raise funding. What are some of the upsides and downsides to this approach? At what point should a firm seek to become "public" and offer stock as an IPO?

   Issuing stock is the same as seeking investors. However, most stock investors will not be interested until you have at least a short track record of success. This could include one year of increasing sales and notice in the marketplace (trade shows, articles, positive product reviews, etc.). And you need the funding to expand your capabilities to increase sales. This is the point at which one would seek to consider an IPO. However, note that issuing an IPO usually requires assistance from a legal firm that specializes in IPOs.

   The upside to issuing stock:

   - *You are not borrowing money but are obtaining funds from investors. Investors recognize there are inherent risks in the stock market, but they are looking to your management team to grow the company to profitability. You will not have monthly loan payments, nor will you ever be required to pay it back.*

   The downsides to issuing stock:

   - *Your management team is expected to provide transparency in all public reporting and in judicious spending of the investors' money. Failure to do this can result, in extreme cases of fraud, jail time and heavy fines.*
   - *If you, the founder, are the CEO, you will report to a Board of Directors (BOD). The Board has a fiduciary responsibility in safeguarding investor funds and that the company's management is trustworthy. If you lose the trust of the Board, they can fire you. The Board will also ask tough questions of you and your senior staff on operations, capital expenditures, strategic planning, and staying on track.*
   - *Investors and business analysts (i.e., Wall Street) will expect increasing sales and profits each quarter. At some point, investors will seek dividends on their shares of stock. Dividends are quarterly cash payments paid out of company profits.*
   - *As the founder/CEO you no longer own 100% of your company; you now co-own it with investors.*

   So, issuing stock has advantages and disadvantages. It is not for the faint of heart. But the funding that investors provide can put your firm on the map and ensure that your company does not run out of money.

3. You will be launching your new product in six months. How will your company and its product get visibility prior to launch?

   A comprehensive marketing and launch plan are necessary for visibility. And this will require funding. Marketing and by extension sales, can plan for and engage in the following activities:

   - *Local press coverage for new products*
   - *Marketing campaign in appropriate publications and on the web*

- *Strong branding and visibility on social media sites such as Facebook, Twitter, Linked-In, etc.*
- *Visibility at trade shows and conferences*
- *Join the local Chambers of Commerce*
- *Special sales incentives and promotions (especially for existing customers).*

4. Whether your firm is small or large, many product development teams are scattered geographically. For example, headquarters may be in Chicago (engineering and marketing), manufacturing facilities in South Carolina, software developers in India and California, and a second manufacturing plant in Indonesia. This makes teamwork, coordination, and communications more difficult. What are some of your ideas to overcome these obstacles? How will you build a team?

   Each company, depending on its culture and organization, will have some in place protocols for dealing with remote teams. However, here are a few suggestions to help coordinate, communicate, and build team morale:

   Regular team meetings, using MS TEAMS, Zoom, or similar platform will help the team relate. Face-to-face is always better. To accommodate time differences, it may be beneficial to schedule meetings at different times so one group is not always meeting at midnight.

   For each meeting, have one of the team members introduce themselves and provide a brief bio: what do they do in their spare time, how long have they been with the company, where are they located, what is their favorite movie, etc. Team comradery is important.

   It is difficult for a PM or PD to manage a product from long distance. If possible (time and budget), it would be useful for the product team leader to travel to each site, as needed, for visibility and a personal touch.

   Finally, the same rules for in-person meetings apply equally to remote team meetings: prepare and stick to an agenda, be aware of time, cover only the topics needed (i.e., don't get bogged on side issues or minutiae), and use your platform's screen share feature so people can see your spreadsheet, presentation, or drawings.

5. How will we know our product is ready for launch?

   If you belong to an incubator group or VC partnership, they will help advise you. If you are working on your own, these are the signs you're ready:

   - *You have done market research and found a market exist for your product.*
   - *You have beta-tested your product with potential customers (if a hardware product, you have built one or more prototypes).*
   - *If a software app, you have channels available to launch your product.*
   - *If a hardware device, you have a manufacturer lined up for production and distribution channels to deliver your product.*
   - *You have a marketing and advertising campaign ready to go.*
   - *You have an internet prescence with payment options ready.*

6. What can a small firm do if the demand for a product outstrips supply?

   If you are launching a software product, it is doubtful this will be a problem as software applications are simply downloaded.

   If you have a manufactured product, it will be necessary to work with your provider (and in turn their suppliers) to ramp up production. If this is not adequate, a second manufacturer may need to be contracted. In the meantime, high demand with limited supply means you can raise your retail price.

7. How would a patent attorney help a small company?
   A patent attorney can assist an inventor navigate through the complex process of obtaining a patent. An attorney will perform a patent search, ensure the patent application is complete and acceptable for filing. Note that each country has its own rules and process for a patent. You may need to hire an attorney in each country you wish to sell your invention.

8. Who monitors the market for patent infringement?
   One might think the US Patent and Trademark Office (USPTO) monitors the market for patent violators, but this is not the case. It is up to the inventor to watch the marketplace for patent infringement.

Notes

1 Brian O'Connell, Benjamin Curry, Contributing Editor, Forbes, May 2021 https://www.forbes.com/advisor/investing/what-are-angel-investors.

2 Raymond Luk, Hockystick/ACA, August 2019, https://blog.hockeystick.co/2019-u.s.-angel-investment-statistics.

3 Draper University Website, August 2021, https://www.draperuniversity.com/blog/what-is-a-business-incubator.

4 Runge, J., Esq. Legal Zoom, What are the Different Types of Patents, https://www.legalzoom.com/articles/what-are-the-different-types-of-patents.

5 The International Union for the Protection of New Varieties of Plants (UPOV), https://www.upov.int/portal/index.html.en.

6 In Business Info.co.uk, https://www.nibusinessinfo.co.uk/content/advantages-and-disadvantages-getting-patent.

Bibliography

Cross, F. and Miller, R. (2004). *West's Legal Environment of Business*, 5th e. Mason, OH: Thomson South-Western.

Noah, A., Patent Rebel. https://patentrebel.com/why-are-patents-important-advantages-disadvantages-pros-cons. (accessed 27 August 2021).

Upcounsel. https://www.upcounsel.com/how-long-is-a-patent-good-for. (accessed 26 August 2021).

7

Manufacturing the New Product

Manufacturing has seen incredible changes over the last 40 years. This is due to many considerations:

- Foreign competition – in both product quality, product costs, and retail pricing
- More advanced products require more advanced manufacturing processes
- Customer demand is driving efficiency and quality
- New advances in manufacturing technology.

In the 1980s and 1990s, US car manufacturers were losing market share to high-quality cars made in Japan. The Japanese cars also cost less. In fact, everything made in Japan seemed invincible: electronics such as TVs, computers, and stereo systems; heavy equipment corresponding to tractors, pumps, cranes, cars and trucks, turbine-generators for electric power production and many other high-dollar equipment. The United States seemed to be losing ground to Japan on almost all industrial fronts. Economists were extolling the virtues of the Japanese "system" and university professors (Peter Drucker among them) were writing books on Japanese management and manufacturing techniques and wondering if we could "copy" their efficiency and processes. As a result, many factories were laying off workers or closing. Some were attempting to mimic the Japanese way with Kanban, Just-in-Time parts delivery, and Quality Circles. They were seemingly helpless to compete.

The balance of trade with Japan became so large that many Japanese firms recognized they would need to build manufacturing facilities in the US. This served a number of purposes. It saved Japan the shipping costs of all the products being sent to North America, it averted possible trade sanctions (i.e., tariffs), and brought jobs to many communities in the US. This also relieved some of the political heat that US elected officials were feeling from their constituents regarding job losses. Further, it expanded the manufacturing capacity and reach of Japan's industries.

It should also be noted that other Western countries were also trying to compete with Japanese quality and costs.

How did Japan, a small Asian country, command such notice and respect around the world? Had the US and European countries lost their competitive edge? Was innovation in the West at a standstill?

The answers are numerous and somewhat complex. If we look at the history of quality in Japan, it is an interesting story. Before and during World War 2 (WW2), Japan's

Product Development: An Engineer's Guide to Business Considerations, Real-World Product Testing, and Launch, First Edition. David V. Tennant.
© 2022 John Wiley & Sons, Inc. Published 2022 by John Wiley & Sons, Inc.

industrial focus was on military hardware. Consumer products were far down on the priority list and consequently, were of low quality (except for some hand made products). After WW2, Japan's industry was rebuilding and focusing on consumer goods. But due to a lack of quality programs, their products still suffered from poor quality and a reputation of shoddy workmanship. In the 1950s, Japanese cars and other products had a poor reputation for reliability. However, in the early 1950s, two American quality experts gave a series of lectures in Japan. The experts were Dr. Edwards Deming and Joseph Juran. Initially, their concepts of statistical quality control and statistical process control were not fully understood by Japanese engineers. Ultimately, the concepts began to take hold and many CEOs of Japanese companies started to embrace and implement the concepts.

Rather than find problems during inspections at the end of production, the focus was to find the root problem and correct these issues wherever they occur on the production line. Applied to whole factories, this not only produces a higher quality product, but eliminates time and cost throughout the production process. It also reduces the scrapping of parts thereby saving material costs. It took Japan 20 years to implement and increase their products' quality. This is how Japan became a powerhouse in quality products: it resulted in significant sales and market share leading to high profitability for Japan's major corporations – at the expense of U.S. and European manufacturers.

Today, items "made in Japan" have a reputation for reliability and high quality (which brings higher prices) as opposed to their previous reputation for "shoddy workmanship" that existed in the past. Interestingly, many US manufacturers did not embrace Deming's and Juran's concepts. Many CEOs in the US simply relegated product quality to the quality department. Many US (and European) companies were caught completely off-guard when Japan's products started to make huge advances in market share in many industries. This did not happen overnight but was the result of 20 years of dedicated effort by Japan's top companies. In summary, Deming outlined in his 1982 book "Out of the Crisis," (Chapter 3, pp 97–98) the seven deadly sins of corporations that contribute to their failure's and poor quality:

1. Lack of constancy of purpose
2. Emphasis on short-term profits
3. Evaluation by performance, merit rating, or annual review of performance
4. Mobility of management
5. Running a company on visible figures alone
6. Excessive medical costs
7. Excessive costs of warranty work.

Let us also examine a partial history of costs in the US auto industry. In the 1960s, US car makers enjoyed an overwhelming dominance in the US market. Also, the Federal Government was investing heavily in the national road system. The $33 billion Federal Aid Highway Act of 1956 funded regional and interstate roads which enhanced the auto industry.

In 1966, the Big Three (GM, Ford, Chrysler) had a market share of 89.6%, which had decreased to 44.5% in 2014.[1] From 1966 to 1969, net imports increased at an average annual rate of 84%.[2]

Below is a quick summary of turmoil the US auto industry has weathered over the years:

- 1960s: stricter safety standards including seat belts, front head restraints, energy-absorbing steering columns, ignition-key warning systems, anti-theft steering column/transmission locks, side marker lights and padded interiors.
- 1970s: extensive government regulations regarding air emissions led to 5 mph impact bumpers, emission controls, phasing out of leaded gasoline – which impacted engine performance with lower engine compression and less horsepower.
- 1970s and 1980s: The big three automakers staggered acutely during this time. There were many problems with engineering, manufacturing, and marketing. These included problems with the Ford Pinto, the Chevrolet Vega, and the AMC Gremlin. These were the most visible problems, but other models had issues as well. This period in US automotive history is commonly referred to as the malaise era of American auto design.[3] Significant engineering and manufacturing problems plagued US car makers, which led to unfavorable consumer perceptions resulting in poor sales and a sense that US cars were of low quality.
- Auto sales were mauled by the 1973 oil embargo by OPEC – the Organization of Petroleum Exporting Countries, primarily headed by Saudi Arabia. The price of gasoline soared and smaller fuel-efficient cars from foreign automakers became very popular and increased their market share of the US market. Further, the Federal Government implemented the Corporate Average Fuel Economy standards (effective 1978). In the 1980s, the economy was weak which affected auto sales and Chrysler Corporation received a federal bailout to the tune of $1.5 billion in loan guarantees (which was later paid back).
- The Big Three began developing joint manufacturing facilities with several of the Japanese automakers. Ford invested in Mazda, Chrysler bought stock in Mitsubishi Motors and GM invested in Suzuki and Isuzu Motors, GM also built a joint manufacturing facility with Toyota known as NUMMI (New United Motor Manufacturing, Inc.).[4]

Despite the economic, regulatory, political, and marketing turmoil, there were numerous technical innovations that helped the auto industry more easily meet safety and environmental regulations. These included disc brakes (instead of drum brakes), fuel injection (instead of carburation), and electronic ignition/engine controls (instead of point distributors). Front-wheel drive became the standard drive train which helped improve fuel economy.

What does this have to do with product development? The above issues illustrate that external factors can impact a company's, technology, market, revenues, and brand loyalty. Think about the inventor or entrepreneur that came up with fuel injection in the mid-1980s. This was a new solution (product or system) that solved a serious problem for all automakers, foreign and domestic. This also goes for electronic control modules. Many times, companies will seek outside experts, consultants, or manufacturers to solve specific problems. Even in times of turmoil, there are opportunities for small companies or entrepreneurs.

Also consider the foreign manufacturers who have set up manufacturing plants in the U.S.: Nissan is in Tennessee as is Volkswagen near Chattanooga, BMW in South Carolina, Mercedes in Alabama, KIA in Georgia, Honda also in Alabama and the list goes on. Surrounding these manufacturing facilities are supplier facilities, creating thousands of jobs and pumping dollars into the local economies, not to mention tax revenues. BMW is exporting some of their SUV models overseas from South Carolina.

Manufacturing still plays a vital role in the economies of many nations. The auto industry is a prime example as everyone understands the car: everyone owns or aspires to own one.

Similarly, there are many other highly capital-intensive industries including aviation (Boeing, Airbus, Gulfstream, Bombardier, etc.), utilities (reactors, wind turbines, etc.) paper and boxboard (boilers, pumps, valves, etc.). Many smaller companies design, build and manufacture many of the components that are used by these larger manufacturers: tires for landing gear, avionics for cockpit instruments, cable for electrical signals in the utility industry etc. These are all potential customers (and their suppliers) for new products and services.

Further, regulations have had a huge impact on many industries. For example, the utility industry has been shifting from coal to cleaner power sources such as wind and solar. New products have been developed to support this industry in its transition.

The State of Manufacturing

Manufacturing in the US has been declining for the past 50 years. However, it is still a large part of the economy providing employment for 12.8 million people with a total output of $2.33 trillion dollars.[5] Manufacturing accounts for approximately 48% of US exports.

If you wish to manufacture a product, there are over 250,000 manufacturing facilities. And manufacturing is getting more efficient and cost-effective with time. Many years ago, numeric control machines were used to manufacture specialized parts in many industries. The machines were programmed (computer controlled) to mill or drill key parts, or to shape metals. This was a great step forward in efficiency as these types of actions would have taken humans much longer. Further, factory automation has also seen acceptance. If one considers the power industry: taking fuel in and producing electricity out, automation has played a large role in plant efficiency. For example, automated systems measure the amount of fuel in (say natural gas) and can adjust pumps, valves, and other equipment to match the demand (running at 25%, 50%, or 100%). This same technology has been used in the paper-pulp industry, building automation, and factory automation.

However, there can be a disconnect between entrepreneurs and manufacturing. Many times, graduate students in science or engineering can develop new products or technologies from their research. They understand scientific principles, but many times they do not understand manufacturing, or the economics associated with production. And just because there are 250,000 manufacturers in the country does not mean each of these has the expertise or the specialized facilities needed for a new product.

If the entrepreneur decides to start a small manufacturing company, where does one start? What are some of the tips for a start up facility? Where can special expertise be found?

As mentioned in Chapter 6, incubators can assist in connecting inventors with appropriate contacts in industry or business. Also, most state governments have an Office of Economic Development which exists to help spur job growth through investments. These offices also foster trade or encourage the installation of manufacturing facilities. These are excellent resources to obtain the appropriate contacts. Finally, there are many trade associations that can also assist.

Some of these include:

American Small Manufacturers Coalition:
http://www.smallmanufacturers.org/who-we-are
National Association of Manufacturers (NAM):
https://www.nam.org
Manufacturing Leadership Council:
https://www.manufacturingleadershipcouncil.com
IQS Directory (Industrial Quick Search Manufacturer Directory):
https://www.iqsdirectory.com/associations/manufacturing-associations.html
IQS allows one to search by category (Automation, Forging, Material Handling, etc.) for dozens of industrial and manufacturing associations.

Thinking About Starting Your Own Manufacturing Company?

There are many considerations for a manufacturing startup as well as some operational issues:

- Some entrepreneurs may consider their product or process highly proprietary (or intellectual property, IP) and do not wish to share this information.
- If well-funded, a new facility can be established, but a consulting engineering firm and/or construction contractor will be needed (sharing IP information)
- It may be worthwhile to purchase a small manufacturer. The machines, and the people who operate them, are already trained and in place.

Inventory

Adequate supplies and materials need to be on hand, including spare parts for machine maintenance. However, investing in too much inventory ties up company funds and space in the warehouse. Consider a just-in-time system or smaller quantities of stock. Do you really need 60 days' supply of nuts and bolts, or will 30 days be adequate?

Safety

There are many local and national safety requirements for manufacturing facilities. Generally, there are state and local fire codes, insurance requirements (Factory Mutual), the national electric code, and OSHA (Occupational Safety and Health Administration). While beyond the scope of this work, safety should be a priority in any factory or production facility. A serious accident or death can bring lawsuits and temporary closings while an investigation is launched. This can also lead to fines if a firm is found to be in violation. However, the real reason to have a safety program is to ensure employees are safe and feel comfortable in their workplace. Regular safety meetings, safety policies, and metrics are highly advisable.

Housekeeping

Keeping your factory clean is also a safety issue. Evidence suggests that those facilities that keep a tidy factory floor have less accidents. It also makes the work environment more pleasant for employees. Dedicate the last 15 minutes of each day to cleaning up.

Note that OSHA has guidelines and requirements for good housekeeping within production facilities. Good housekeeping boosts employee morale, improves quality control, and helps maintain better control over the production process.

Factory Layout – Process

A factory takes in raw materials at one end and a finished product is shipped out at the other end. Some factories take in finished components and assemble them into a final product. These are known as assembly plants. How the factory is arranged is important for efficiency and cost savings. It makes sense to have material and parts flow where the assembly process occurs sequentially from station to station. However, over time, new production lines can be installed, or a "custom fab" area for special orders is set aside – all of which means more shuttling of parts and less efficiency. Having to cart parts back and forth from one end of the facility to the other is not cost efficient.

If we consider our Air Emissions Monitor from previous chapters, Figure 7.1 Manufacturing Flow, shows a hypothetical flow for production of the new air monitor from start to finish. Notice that some components arrive already assembled, while some are fabricated in place. Figure 7.1 is one possible flow scenario. There may be other floor arrangements that are more effective. Generally, industrial and/or manufacturing engineers compose various floor arrangements to arrive at the optimum workflow.

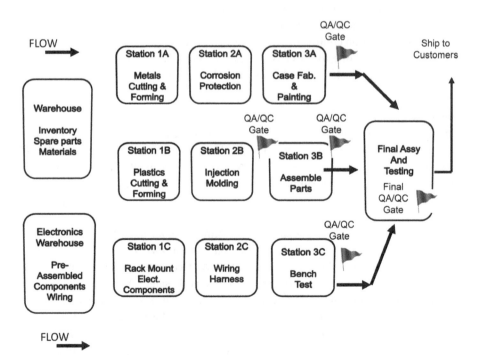

Figure 7.1 Manufacturing Flow. Figure developed by David Tennant

Notice there are several Quality Control (QC) checks within each manufacturing line. This means there may be some parts that do not meet standards and may be rejected to a waste bin. The manufacturing and engineering team should determine the rejection rates and root cause of any problems. Waste parts and materials represent lost costs and wasted time. QC also ensures that the individual components are working, and that the assembled final product is working. This ensures customer satisfaction and reduction of returns or warranty repairs.

Notice in Figure 7.1 that there are three work flows in this facility: metals (top), plastics (center), and electronics (bottom). These are occurring simultaneously, and each flow needs to have the correct number of finished parts available for final assembly at the same time. Too many wiring harnesses produced, compared to the steel and plastic parts, will mean lost optimized time and excess wiring "waiting" for final use ... it may begin to back up, slowing down the electronics line. This means each line must produce the right number of parts at the right time; otherwise, some lines will begin to back up for being too far ahead while others will be behind. Industrial engineers usually perform time motion studies of worker productivity and can determine with some accuracy how to staff each line and how long each workstation will need to produce their parts. Further, robots and smart factories are beginning to eliminate bottlenecks such as these and ensure the smooth flow of materials and tasks. More on smart factories further on.

Supply Chain Management

While details of Supply Chain have been offered in previous chapters, it is advantageous for a small manufacturer to have a trusting and long-term relationship with suppliers for the following reasons:

- Some suppliers may be willing to stock parts or materials at their warehouse or distribution point thereby reducing the inventory held in stock by you.
- A strong relationship with your suppliers means they may be more flexible in payment terms and willing to go the extra mile when materials need to be expedited or obtained on an emergency basis.

It is necessary that a manufacturer qualifies suppliers. It would not be productive to enter into a contract with a company experiencing cash flow problems or approaching bankruptcy.

Small vs. Large Manufacturers

A small manufacturer offers several advantages over a larger one. For example, generally they can be set up more quickly; that is, they are nimbler in changing operations. This might involve changing or reprogramming numerical control (NC) machines or designing and fabricating tooling. They are more likely to be efficient for small batch runs as opposed to continuous operations.

If a large run (say 100,000 products) is needed, then most likely a larger manufacturer will be better suited.

Some of the advantages of small producers include:

- Closer management of the production process
- Nimbler to change when product demand changes or for custom orders

- Less capital dollars needed
- Closer supervision of workers
- Less possibility of disruption due to labor or supply issues.

Disadvantages of smaller production facilities:

- Less robotics – it may take longer to produce a product
- Less consistency or lack of quality – directly related to lack of robots. Whenever people are producing a product, they cannot be as precise or as quick as robots
- Generally, more waste due to rejected or substandard parts
- More susceptible to economic disruption. Can a smaller firm survive an economic downturn?
- Older facilities. Smaller producers generally do not have the capital to invest in new, high-tech machines or robots.

Just-In-Time Manufacturing (JIT)

Just-in-time means materials or components are delivered when needed and are not stored long-term in the warehouse. It is a workflow method focused on reducing production flow times, and response times from suppliers. If is a "pull" approach meaning that production is matched to demand. Many times, companies will build a quantity based on expected demand, which may not be accurate. If a facility can make batch runs: that is, smaller units in a single run, this will allow all units to be sent to market. No excess units in storage or excessive inventories and parts in the warehouse.

JIT is a system and not just a method of inventory reduction. JIT has been around for at least 60 years and was started in Japan. At that time, factories could not find the necessary skills needed for production or the large capital required to compete against more developed countries.

As a result, JIT is a system using smaller factories, smaller batch runs, and other techniques. Below are some of the benefits of a JIT approach to production:

- Small batch production runs
- Skill diversification: workers that can perform multiple roles or tasks. This means that workers can be shifted to where critical operations are needed
- Eliminating defects – more focus on quality and wastes minimization. It is possible that robotics can assist in this area (minimal human error)
- Consistent plant load – less swings in production
- The efficient movement of materials from beginning to end. This includes suppliers, parts movement from the warehouse to the factory, and movement of materials between workstations
- Less inventory. Due to smaller batch runs and more frequent materials deliveries, there is less need for material storage or storage space
- Less space required to operate. Smaller batch runs will mean fewer machines, smaller facility space, and optimizing worker skills for different tasks.

JIT offers reduced production time and operating costs; and faster product to market. Like any system, there are inherent risks which can be devasting for a small company if these risks are not identified in advance and minimized. For example, if a critical machine breaks down during a production run, it can delay product shipment to customers and, potentially, shut down the whole production line. This means facilities need to have a strong preventive maintenance program so that all critical machinery is kept in top condition.

Further, having suppliers close by is important as shipment of raw materials or components should adjust to production schedules and quantities. If a supplier is thousands of miles away, this will not be conducive to smaller, more frequent deliveries.

How is Lean Manufacturing Different From JIT?

The terms lean manufacturing and JIT manufacturing are often used interchangeably. However, there is a distinct difference. JIT, as noted, seeks to streamline factory production and the supply chain process to reduce costs.

Further, JIT is focused to deliver one item (a batch run) with minimal flaws or variations. JIT provides more flexibility, whether to produce larger or smaller product amounts, it will be more in tune with market demands including fluctuations. As a result, manufacturing equipment needs to be adaptable and easily changed to satisfy new or different products.

Lean manufacturing, on the other hand, is more concerned with the process outside of the factory and focuses on the customer. Specifically, this means determining what the customer wants. Collaborating with customers to determine wants and needs can lead to a more focused design. Therefore, functional areas, such as engineering, sales, and marketing may have extensive discussions or working sessions with customers. Recall in previous chapters, discussions on the importance of customer focus groups and beta testing.

Onshore vs Offshore

Manufacturing employment in the US declined by 5.8 million jobs between 2000 and 2010.[6] A significant number of those jobs went to countries in Asia. This is because labor and production costs are extremely low compared to the US and other Western nations.

It is also known that when manufacturing is exported to other countries, the expertise and future innovation usually goes with those production techniques. There are also issues with preserving intellectual property and patents. Therefore, an inventor in search of a manufacturer must balance the cost of production with protecting processes and patents. Further, managing product quality can also be problematic from half a world away.

There are advantages and disadvantages for moving production offshore. It allows companies to utilize their core competencies. For example, a US firm may have creative and innovative people, great engineering expertise, and strong marketing prowess. The foreign firm will most likely offer abundant and low-cost labor, thereby using its core competency. When a domestic company decides to outsource its engineering and innovation (downsizing to save costs), and entirely removing a key part of the organization, this is where issues can arise. It is difficult to control quality and supply chain problems on the opposite side of the planet. Companies need to carefully craft a business strategy for offshore production.

It is recognized that competitive issues are at play. Higher production costs will mean a higher retail price for a new product. Therefore, a careful review of costs and benefits must be considered. However, there are new and interesting advances that are beginning to appear in manufacturing facilities and has the potential to bring further competitiveness to domestic manufacturing.

New Manufacturing Advances

3D Printing

3D printing technology is becoming more accessible and offers a variety of advantages. First, molds are not required, therefore the costs associated with this process can be very competitive. Further, the output is based on CAD or STL files and product customization can be accomplished without incurring additional manufacturing costs. 3D printing has been around for over 30 years, but over time, it has become more adaptable and cost effective.

The technology can produce prototypes quickly which helps bring products to market more quickly. Cost savings can also be realized with less waste. Only the exact amount of material is used, as opposed to previous manufacturing techniques requiring cutting, which will always produce waste material.

Not all materials can be used. The most common materials for 3D printing include plastics, metals, resins, and ceramics. The printer contains a spool of plastic (or metal) that is produced one layer at a time until the product is completed. This provides design flexibility, and the machine can be left alone (no operators required) while it completes the tasks (usually in hours, not days).

Figure 7.2   3D Printing. NEW DATA SERVICES / Unsplash

Disadvantages to 3D printing include high cost with high volume; economies of scale do not occur with 3D in comparison with CNC machines and injection molding. Figure 7.2 shows a small 3D manufacturing printer.

Robotics

Complex new products will require advances in manufacturing technology.

Robotics is a part of this advancement. Robots on the manufacturing line are faster and more precise than a human doing the same work. Further, this provides consistent parts quality, which in turn saves time, minimizes waste, and reduces costs. Consistency in parts and products will ensure higher quality and reduce warranty work – saving costs and company reputations. Robots for industrial use have been available for some time, but they are becoming more advanced as higher precision is required. Early robots were not "smart" and could not function without human interaction. The tasks they performed were monotonous and had to be programmed.

The next generation of robots were controlled by programmable controllers making them smarter, autonomous, and less reliant on human manipulation. Modern robots can control the flow of materials and manage production tasks. As time moves forward, it is likely that manufacturing facilities will employ more robots than people. Besides drilling or forming materials, robots also pick up completed materials from one machine and place into another machine for the next task (using the next robot). Further, robots can also perform spot welding, grinding, spray painting and polishing. Some of these tasks are guided by lasers and are necessary for highly complex tasks.

It is worthy to note that robots can also perform tasks that are dangerous or toxic to humans. For example, after the Fukushima Daiichi nuclear plant disaster in 2011, radiation levels were too dangerous to allow humans to survey the reactor damage. Consequently, robots were used to inspect and provide video images of the damaged power plant.

Figure 7.3   Manufacturing Robotics. Lenny Kuhne / Unsplash

Robots today and in the future are collecting data from the "cloud" and can analyze sensor data in real time, allowing them to change direction of parts or material flows. Artificial intelligence (AI) is allowing robots to be smarter, more autonomous, and efficient than humans could ever be.

The use of robots in industry has been increasing significantly. The International Federation of Robotics (IFR) states the global average for industrial robots per 10,000 manufacturing workers grew from 66 in 2015 to 85 in 2017.[7] Figure 7.3 shows a robot at an auto manufacturing plant.

Robotics and the Use of Internet of Things (IOT) and 5G

In Chapter 8, IOT is listed as a new technology from the standpoint of micro grids and smart cities. IOT has also found applications in manufacturing technology, especially robots. IOT and 5G (also in Ch. 8) will allow the integration of data, operations, people, and equipment in real time. Robots will be able to communicate with each other and, using artificial intelligence, sort through data (collected by multiple plant sensors) and make operational adjustments – that is, make decisions. Artificial intelligence will increase the use and functionality of robots. The future for robots in manufacturing is exceptional. Note that modern production facilities are highly automated and employ sensors in many locations across the plant.

Sensors in manufacturing facilities are used to measure temperatures, pressures, flow rates (fluids or gases), and other media. The data from sensors can be utilized in turn to control fluid flows, temperatures, etc. by modulating the performance of compressors, motors, valves, pumps, and other process equipment. Why not also coordinate robots so that they can communicate with the sensors and each other? The advantages to 5G are that higher volumes of data (throughput) at much faster speeds will occur and this will have a large impact on future factory automation.

It should be apparent that the joining of smart robots, factory automation, 5G, the cloud, and artificial intelligence will usher in a revolution in manufacturing technology and practical applications. This will allow the production of new products faster, cheaper, and with higher quality and consistency. Products made in smart manufacturing facilities will have better quality and consistency than anything "hand-made."

Predictive or Preventive Maintenance

Maintenance in factories has always been a challenge. Some facilities shut down for two weeks every year to perform plant upgrades and maintenance on highly critical equipment. This is known as planned maintenance or a planned outage. It can be costly and damaging for a critical machine, pump, etc. to break down during a production or batch run. It is entirely possible that one machine failure can cause other machines, upstream or downstream, in the production process to also fail, not to mention that some materials or parts may have to be scrapped. This would be characterized as a forced or unplanned outage.

Consequently, production facilities cannot wait for machines to fail before fixing or replacing them. The whole purpose of preventative maintenance is to *prevent* machines from breaking down by performing maintenance on a regular schedule. For example, if we know that a critical valve will function for 18 months before the packing fails, we can schedule a work ticket every 15 months to prevent the valve from failing. This is also a similar approach for other equipment such as motors, pumps, etc.

However, smart factories, sensors, equipment monitors, and other data will enable the robotics or control system to notify plant personnel when equipment performance is starting to lag or act suspicious. These new technologies will have a huge impact on keeping plants running for longer periods and with better predictive/preventive maintenance. With intelligent planning, it may be that fewer spare parts are stored in the warehouse thereby reducing inventory, saving costs, and reducing space requirements.

Robotics in New Product Development

From a product development perspective, advanced manufacturing looks highly promising. However, not all factories employ sophisticated robots due to the capital costs involved. Most complex robotic systems are likely found in the larger, more advanced factories (planes, autos, etc.) in the global marketplace.

It is important to note that robots are excellent at batch runs but need to be reprogrammed for each individual batch. This involves time for programming and several test runs to confirm the programming is correct. Further, does every new product need the sophistication of robots? Probably not. Smaller manufacturers many times employ robots on a limited basis (not necessarily "smart") and this may be acceptable for many small company products, limited batch runs or entrepreneurs producing their first product.

Chapter Key Points

- Manufacturing has been under pressure for the last 50 years to be more competitive and cost effective.
- Many factors can impact manufacturing including, but not limited to: foreign competition, government policies and regulations, monetary policy (i.e., interest rates), currency exchange rates, supplier issues, and disruptive new technologies.
- In the US, manufacturing plays a large role by employing almost 13 million people and helping the balance of trade by exporting almost half of the country's manufactured goods.
- Manufacturing accounts for $2.3 trillion in total output.
- For small companies looking to produce a product, it may be beneficial to seek smaller companies that possess the expertise or buy an existing small manufacturer.
- Key ingredients to cost effective production include inventory management, a strong safety program, good housekeeping, and an effective plant process layout.
- Just-in-time manufacturing is concerned with activities in the plant warehouse and within the manufacturing facility. Lean manufacturing is focused on customer needs and is external to the manufacturing facility.
- New advances in factory automation, robotics, 5G, IOT, and other technologies will have a huge impact on the future of manufacturing.

Discussion Questions

1. Why is housekeeping at a factory considered a safety issue?
2. Figure 7.1 shows a proposed manufacturing flow to produce an air emissions monitor. Looking at this diagram, which production functions could potentially be performed by robots?

3. Referring again to Figure 7.1, what advantages would robots bring to this facility?
4. Are there any disadvantages to just-in-time (JIT) manufacturing?
5. What are some of the advantages and disadvantages of offshore manufacturing?
6. If your company is building a new state-of-the-art factory, what are some of the considerations (pro and con) in determining its location?

Discussion Question Answers

1. Why is housekeeping at a factory considered a safety issue?
   Good housekeeping means clean floors, tidy work areas, and minimal trash or materials laying around. Research has shown this translates into fewer accidents and sets the tone for the corporate culture. Further, OSHA has guidelines and requirements for good housekeeping within production facilities. Good housekeeping boosts employee morale, improves quality control, and helps maintain better control over the production process.

2. Figure 7.1 shows a proposed manufacturing flow to produce an air emissions monitor. Looking at this diagram, which production functions could potentially be performed by robots?
   It is likely that workstations 1A, 2A, 3A, 1B, 2B, and 1C could utilize robots for these functions. It would depend on whether the robots are "smart" as to how much production could be performed by robots.

3. Referring again to Figure 7.1, what advantages would robots bring to this facility?

   - *Better quality and consistency of individual and assembled components*
   - *Less scrap or wasted materials*
   - *Less warranty work or retuned products*
   - *Can work more precisely and faster than humans*
   - *Cost savings based on all the above.*

4. Are there any disadvantages to just-in-time (JIT) manufacturing?
   JIT works very well for most production facilities. However, if suppliers run into trouble (labor strike, unable to procure materials, etc.), this will have an immediate effect on production as JIT does not encourage storage of parts or materials in the warehouse. JIT uses customer orders or demand to start production. Any disruption in the supply chain will disrupt product output and delivery.

   Also, JIT requires strong coordination during the manufacturing process. This means that companies must make an investment in supply chain IT so that materials can be automatically ordered when stocks run low.

5. What are some of the advantages and disadvantages of offshore manufacturing?
   Disadvantages

   - *Potential loss of intellectual property*
   - *Transferring manufacturing overseas usually means that country will benefit from future manufacturing advances*
   - *Some lower end jobs will be lost in domestic facilities (i.e., transferred offshore)*
   - *More difficult to control quality and information flow*
   - *Disruptions in supply chain harder to mitigate.*

Advantages

- *Lower labor costs in offshore country*
- *Comparative advantage: for example, domestic firm has expertise in engineering, creativity, and marketing; offshore firm in low-cost labor. This allows both companies to utilize their core competencies.*

6. If your company is building a new state-of-the-art factory, what are some of the considerations in determining its future location?

- *How close are suppliers?*
- *Are there enough 5G towers nearby to implement IOT?*
- *Is the facility near a major port or airport?*
- *How good is the education system (well educated workforce)?*
- *Are there adequate and reliable power and water resources?*
- *How strong is the local infrastructure (roads, communications, etc.)?*

Notes

1 Joel C., Brooks, D., Mulloy, M., Economic Policy Institute (May 6, 2015), The Decline and Resurgence of the US Auto Industry, accessed on August 25, 2021. https://www. epi.org/publication/the-decline-and-resurgence-of-the-u-s-auto-industry.

2 Okubo, S., "Foreign Automobile Sales in the United States," in Federal Reserve Bank of Richmond. "November 1970," Economic Quarterly (Federal Reserve Bank of Richmond) (November 1970), accessed on August 25, 2021. https://fraser.stlouisfed. org/title/960/item/37804/toc/174839.

3 Martin, M. The Truth About Cars, May 5, 2001. Accessed on August 25, 2021. https:// www.thetruthaboutcars.com/2011/05/ what-about-the-malaise-era-more-specifically-what-about-this-1979-ford-granada.

4 "In Partnership, the United States and Japan 1951-2001", ed. by Akira Irike and Robert Wampler.

5 National Association of Manufacturers, accessed August 30, 2021, https://www.nam. org/state-manufacturing-data/2020-united-states-manufacturing-facts.

6 Bonvillian, W. and Singer, P., Advanced Manufacturing, MIT Press, Cambridge, MA., 2017.

7 Atkinson, R., ITIF, Information Technology and Innovation Foundation, accessed September 10, 2021, https://itif.org/publications/2019/10/15/ robotics-and-future-production-and-work.

Bibliography

https://www.fingent.com/blog/top-10-technologies-that-will-transform-manufacturing-in–2021.

https://waypointrobotics.com/blog/manufacturing-trends.

https://www.arnoldmachine.com/6-exciting-advances-manufacturing-automation.

https://www.twi-global.com/technical-knowledge/faqs/what-is-3d-printing/ pros-and-cons#Prosof3DPrinting.

https://maycointernational.com/blog/how-advanced-technology-is-changing-in-the-manufacturing-industry.

8

Engineering Product Design and Testing

The profession of engineering uses the principles of science and mathematics to solve technical problems. However, in any engineering engagement, this also includes creativity, vision, perseverance, and compliance with federal and state laws. For example, this could include environmental regulations (e.g., air and water emissions) during the manufacturing process.

Managing the Approved Scope and Budget – Why Is This Important?

Recall that the Work Breakdown Structure (WBS) defines the scope of our project. And, Product Development can be considered a project. It is important to recognize that some firms reinvest around five to seven percent of their profits back into R&D (i.e., new product development). The author is aware of one Fortune 100 company returning 11 percent of its profits to R&D. However, the budgets for new product development are not unlimited. This can also be more significant for publicly held companies. A publicly held company is one that has its stock listed on the stock exchange (i.e., NASDAQ, NYSE, etc.). The owners of the company are the shareholders who have invested their money in the firm. Shareholders expect the value of the company to go up over time; and they generally like to see some of the profits returned to shareholders in the form of dividends (quarterly cash payouts). This is opposite privately held companies that are either family owned, or employee owned. Tables 8.1 and 8.2 indicate a few publicly and privately held companies.

Publicly held companies must follow federal guidelines and regulations related to accounting transparency. Privately held companies have less strict rules as the owners are not the public.

As noted in Chapter 2, the business case is a detailed proposal for a new project, product, or business. The timelines and costs in a business case are a best-case estimate. Many times, these estimates will be provided as your working budget, much to the dismay of product teams. Note that an estimate is an educated guess as to what the project will cost. A budget is what you are allowed to spend. It is necessary to confirm that the budget is realistic by the project/product team.

Product Development: An Engineer's Guide to Business Considerations, Real-World Product Testing, and Launch, First Edition. David V. Tennant.
© 2022 John Wiley & Sons, Inc. Published 2022 by John Wiley & Sons, Inc.

Table 8.1 Sample of Large Publicly Held Companies.

Company (ticker)	Products	Revenue (billions)	No. Employees
Coca-Cola (KO)	Beverages, non-alcoholic	$ 294.1	80,300
Apple (APPL)	Consumer electronics	$ 267.7	147,000
Bank of America (BAC)	Banking, financial services	$ 74.2	213,000
Wal-Mart (WMT)	Discount stores	$ 559.2	2,300,000
Amazon (AMZN)	Internet retail sales	$ 386.1	1, 298,000
Volkswagen (VWAPY)	Auto manufacturing	$ 262.3	662,575

Data from finance.yahoo.com (4/5/2021)

Table 8.2 Sample of Large Privately Held Companies.

Company	Products	Revenue (Billions)	No. Employees
Koch Industries	Chemicals, Consumer Products	$ 115.00	100,000
Bechtel	Engineering and Construction	$ 21.80	55,000
Cox Enterprises	Media	$ 21.10	50,000
Cargill	Food, Drink, Tobacco	$ 114.60	155,000
Price Waterhouse Coopers	Business Services	$ 43.00	276,000
Fidelity Investments	Financial Services	$ 20.90	50,000

Source: Data from Forbes, 2020 Rankings, viewed February 27, 2021, https://www.forbes.com/largest-private-companies/list/#tab:rank

The Project Lifecycle

All activities within a company can be considered a project. A new marketing campaign, a new product, building a new hospital, etc. are all examples of projects. It is only in the last 20 years or so that companies have recognized having a robust process for managing projects gives them a competitive advantage. How much better off financially would your company be if 90% of projects came in on budget and schedule? Project management is a strong tool that can help companies more effectively design and develop new products and services. Note that an awareness of budget, schedule, risks, etc. is monitored and controlled throughout the life of the project.

Figure 8.1 shows a sample project lifecycle (do not confuse this with the product lifecycle). This is very typical, but companies may have their own approach.

Many companies have designed processes and procedures that align with the project lifecycle.

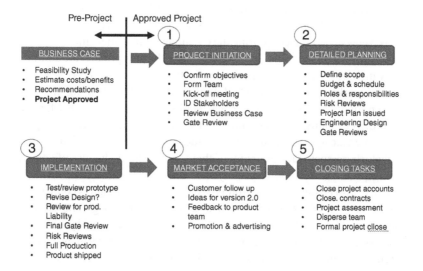

Figure 8.1 Sample Project Life Cycle. Figure developed by David Tennant

The business case, discussed in Chapter 2, is generally developed prior to formal project launch. Remember, it is a feasibility study produced to help corporate executives make a go or no-go decision.

Initiation – Once the product or project is approved and funding secured, the project moves into the initiation phase. Primarily, this is where the project leader (or product leader) identifies key stakeholders, and the sponsoring executive formally launches the project by issuing a memo or project charter. This will identify the project's existence and the person who will lead the effort. It is appropriate for the team to have a kick-off meeting to go over the project's objectives and confirm each team member's role and responsibility. Some companies may also have preliminary schedule and budget discussions with the team.

Planning – This is where detailed design and planning will occur. Some of these activities include (but are not limited to) are shown in Table 8.3:

Table 8.3 Project Planning Activities.

• Develop communications plan	• Review company procurement policies and procedures	• Determine how/when risk reviews will occur
• Reaffirm product objectives	• Develop prototype QA/QC testing procedures	• Meet with key stakeholders to obtain input, desired features, etc.
• Confirm budget and schedule are realistic	• Determine engineering design review points (Gates?)	• Review staffing requirements
• Meet with production: logistics and costs	• Scope change management	• Define project success

Table by David Tennant

The planning phase is critical and should have ample time scheduled for these activities. Many times, companies rush through the planning stage, only to realize items that were not fully planned show up as problems during the execution or implementation stage.

Implementation – Implementation is where and how the new product is manufactured. This may include the addition of new manufacturing lines, new processes, and training of skilled workers to ensure product quality. During the planning phase, it should have been identified that new processes, equipment, etc. are needed for the project to succeed. The quality assurance and quality control group (QA/QC) will play an integral role in ensuring product quality. Production engineers will identify and correct manufacturing issues. Ideally, many of these issues should be identified and corrected during prototype testing. A prototype is a first run and test of the new product. There may also be iterations of prototypes before a final design is confirmed.

Continuous Review – It will be necessary to manage the schedule, budget and activities throughout the project's lifecycle. If the new product goes way over budget, and by extension, schedule, it may make the new product uneconomical to produce or too costly in the marketplace. Remember, a company cannot stay in business unless it makes a return on its investment.

Another way to monitor the project is to perform risk reviews on a regular basis. New potential problems will emerge throughout the project and the project team needs to identify these in advance so that mitigation strategies can be developed.

Project End – The finish line is a combination of actions that signify the project is at an end. Some of these actions include successful production and shipment of new product to the marketplace; achieving all of the objectives identified in the project plan, closing accounts, and performing a project assessment.

All the previous activities will require exceptional management and technical talent to be successful. Remember that open communications are necessary with stakeholders, executives, functional departments, and suppliers.

A True Story: Ignoring the Warning Signs

A large telecommunications company was investing significant dollars in one of its manufacturing facilities. The Marketing group had made strong projections for future sales of the firm's new product – currently under development. Capital spending was provided to increase the number of manufacturing lines, add new material warehouses, increase R&D spending, and install a new clean lab.

Management had placed ambitious goals for the product's release and the teams were putting in a lot of overtime to reach the deadline. As a result of the expected demand, and to also beat their competitors to market, there were many rushes through R&D, scope changes were becoming a problem, and many large purchases did not go to competitive bid (to save time). As a result, the program went significantly over budget and behind schedule. Finger pointing was occurring between different factions at the corporate and plant levels. There were many problems that may have been uncovered with a risk review:

- Not going out for bid meant that pricing for expensive capital equipment was high, absent any competitive pressures.
- R&D, Marketing, Engineering, and Production had not worked together to develop a timeline. The schedule was dictated by senior management based on "expected" moves by competitors. For this reason, the schedule was unrealistic.

Figure 8.2   R&D Lab. ThisisengineeringRAE / Unsplash

- The budget was tied to schedule milestones. Since the schedule was not valid, neither was the budget or cash flow projections.
- The initial problems in developing the new product were not taken seriously, which came to fruition during the prototype manufacturing stage (end of planning but prior to implementation phase).
- Several key people were looking to "bail out" of the project, sensing that failure may be on the horizon.
- It was discovered, late in planning, that several environmental issues would need to be addressed. This added further costs to the budget and additional schedule delays.
- Marketing and Engineering argued over changes to the original scope. Engineering was trying to hold to the agreed-to scope and cost. Marketing was pushing for changes to the product that would make it more desirable in the marketplace. Changes to the product meant that R&D and Engineering would have to redesign and test the newly revised product. These actions had serious budget and schedule implications.

A formal risk review would have uncovered many of these items earlier in the project thereby saving time and money. Let's move on to the actual actions needed to perform a risk review. Figure 8.2 shows a typical R&D lab, where many initial problems begin.

Preventing Failure and Surprises: Performing a Risk Review

One of the most useful tools for anticipating problems, and thwarting them, is a risk review. Put simply, a risk is a potential future problem. If the product team can identify problems likely to occur in the future, they can devise strategies to avoid or minimize them.

First of all, risk reviews should be conducted throughout the project lifecycle; that is even starting with the business case running through the end of the project. This is because new problems will emerge as the project develops (and some will fade out).

Key Tenets of Risk Reviews

- Risk reviews are a team effort, led by the team leader or a facilitator. Figure 8.3 illustrates a team risk review session.
- Risks can only happen in the future.
- Risk identification is typically done as a brain-storming session.
- Risks can be interdependent. For example, a problem triggered in Dept A may lead to issues downstream in Dept B.
- Performing risk reviews will make you a better manager and the team more aware of potential snags in the project.
- Identifying risks and developing strategies will minimize the chaos and "firefighting" that sometimes occurs in projects.
- Finally, the ultimate goal is to change today's actions to improve future results.

Two Types of Risk Review: Qualitative and Quantitative

There are two types of risk reviews. The most common approach is the qualitative review. Table 8.4 is an actual risk review using the qualitative method.

Notice that for each risk, we have assigned a Probability (P) and Impact (I). As a team, we must decide what is the probability of the risk occurring (High, Medium, or Low); and, if the risk occurs, what is the impact (High, Medium, or Low). This can be somewhat subjective but relies on the collective judgment of the team or subject matter experts (SMEs). Also, for each risk, there are several strategies that have been devised. One rule of risk reviews is that there should always be two or more strategies for each risk. Further, some of these strategies may require schedule or budget revisions. This is especially important for the troublesome field instruments (item 4). If these are to be replaced, then this activity should be placed on the project schedule and additional dollars budgeted for the new hardware.

This illustrates the usefulness of a risk review. In the above example, we are anticipating problems and can address these in advance rather than firefighting when they appear in the implementation stage.

Figure 8.3 Risk Reviews are a Team Activity. Jason Goodman / Unsplash

Table 8.4 Qualitative Risk Review of New Data Acquisition System.

	Risk	P	I	Strategies
1.	Supplier may be late in shipping system, thereby slipping project schedule	M	H	• Place early order for system • Have supply chain keep on top of supplier • Assess late penalties (liquidated damages) • Strong communications with supplier on weekly basis
2.	System may fail field tests: potential schedule slip	L	H	• Perform factory testing prior to shipping • Ensure regular design reviews to catch errors • Write performance guarantees in contract
3.	Key technicians may not be available when needed during startup	M	M	• Make arrangements early to schedule techs • Enlist support of executive sponsor (VP) to ensure that supporting techs are available • Have external contractors available as back up
4.	Some field instruments may be a problem, especially those in high vibration or hazardous areas	M	M	• Ensure all field instruments (sensors, controllers, etc.) are inspected and tested prior to start up • Replace known problem instruments prior to system arrival • Double check cables and terminations in wiring cabinets
5.	A severe winter storm could delay testing and start-up activities	M	L	System and site testing occurs in adverse weather on a regular basis. This is not considered a major issue. All systems are indoors.

Table developed by David V. Tennant

The second type of risk review is quantitative, which implies the use of numbers. This is a technique that can be used when we have a good feel for the cost of impacts. Let's look at a simple, single risk using a quantitative approach:

Risk	P	I	EMV
Supplier may be late providing key components in time for implementation	60%	$250,000	$150,000

First, here are some definitions to be aware of:

P – Probability, expressed as a percentage or decimal. For example, 60% = 0.60

I – Impact expressed in dollars (or other currency). In this case, the expected impact is $250,000 if the supplier is late.

Notice that these two numbers are also subjective. What if the probability is closer to 80% or 90%? How do we know it is 60%? How do we know the impact if this risk occurs is $250,000? These numbers are based on the collective judgment of the team.

Finally, know that the EMV or expected monetary value is the product of $(P) \times (I) = $ EMV. In this case, $0.60 \times \$250,000 = \$150,000$

So, how do we use this? If this event occurs, we need $250,000. The sum of $150,000 will not be enough. This concept is more fully illustrated in Table 8.5.

Table 8.5 Quantitative Risk Review.

Risk	P	I	EMV
Supplier is late with system delivery	0.6	$ 600,000	$ 360,000
Shortage of qualified technicians during changeout	0.25	$ 300,000	$ 75,000
Support equipment not available when needed	0.7	$ 200,000	$ 140,000
Testing QA/QC is weak	0.1	$ 150,000	$ 15,000
System training is inadequate	0.2	$ 200,000	$ 40,000
Software has too many bugs prior to change	0.4	$ 300,000	$ 120,000
New facility not ready when needed	0.1	$ 75,000	$ 7,500
IT drops late	0.1	$ 30,000	$ 3,000
Rain prevents transfer of support equip	0.1	$ 25,000	$ 2,500
Snow delays arrival of out-of-town tech's	0.15	$ 50,000	$ 7,500
New system won't fit on elevator	0.3	$ 150,000	$ 45,000
System fails FT	0.3	$ 100,000	$ 30,000
System fails Pre-op field test	0.1	$ 50,000	$ 5,000
System documentation not ready	0.2	$ 40,000	$ 8,000
HVAC system undersized or fails	0.25	$ 65,000	$ 16,250
Unrealistic schedule milestones	0.4	$ 125,000	$ 50,000
New Mgmt. in place with company reorganization	0.8	$ 250,000	$ 200,000
Totals		$ 2,710,000	$ 1,124,750

Table developed by David V. Tennant

There are several items to point out with table 8.5.

The probability of *all* these problems occurring is probably very low, perhaps even 0%. The probability of *half* of them occurring is also low, perhaps 10–15%.

However, it is likely that *some* of them *will* occur. If we look at the total for the EMV column, we note that it equals $1,124,750. We can use this amount to add to our contingency budget. It is likely that this amount will be adequate to address the five or six problems that will develop. This is just one approach. Recall that all projects, including product development projects, should have a contingency budget. Some companies arbitrarily assign a 10, 15, or 20% contingency. However, the quantitative technique is a more analytical approach.

To some extent, this model also is subjective. For example, the first line (supplier is late): how do we know the probability is 60%? Also, how do we know that the impact is $600,000? Do we have previous history to go on? However, note that if we have identified these problems, we can take steps to prevent them. Therefore, we may be able to prevent any of them from occurring.

The author has observed companies use both models (quantitative and qualitative). The key point is to perform risk reviews on a regular basis to identify potential problems and prevent them from occurring. As for engineering, it would make perfect sense for this group to perform their own internal design risk reviews.

Design and Status Reviews

Design reviews are an important and necessary step in the product development process. In engineering, design reviews are key to designing and building a successful outcome. For example, if an engineering firm is designing a new bridge, it is necessary to have review points to ensure the design meets specifications. Specifications can be stated by the client (say, for instance, a state Department of Transportation or Federal agency). Generally, there will always be standards and regulations; and the design must be in compliance with these requirements. In this case, the public may be harmed if the bridge fails. Design reviews should be tied to milestones in the schedule. Some design firms simply hold a review at the 20%, 50%, 70%, and 95% progress points.

For products in development, it makes sense to hold design reviews at points tied to the schedule. This can be gate reviews and other milestones such as deliverables: prior to prototype production, testing, data review points, etc. This can also include preliminary designs and reviews during scope change requests.

Design reviews should always compare the status of the product's design with the requirements provided in the approved scope and objectives. This is important as scope changes along the way may change the functionality of the product. Further, most prototypes will undergo testing to ensure the product is robust enough (i.e., safe) for the market.

As an example, commercial grade aircraft go through significant testing whether a small light aircraft (e.g., Piper, Cessna) or major jetliner such as those built by Boeing. Planes will go through air tunnel test to determine how wind affects the drag or lift of a particular wing. Additionally, wings are also subjected to continuous stresses (bouncing up and down on a shaker) for weeks to determine where the weak points for cracking or material fatigue might appear. Today, there are computer program models to assist with predicting failure points or performance under stress conditions.

Regardless, the takeaway is that product testing is required to determine if a design should be revised or strengthened; or if it is on the right track. Therefore, design reviews are appropriate both before and after testing has occurred and the data have been analyzed.

Note that design changes are much easier and less costly early in the project.

Figure 8.4 shows the relationship between design costs and time. It is less costly to make needed changes during the early phase of a product's development. Once a product is in production, the cost of changes can be prohibitive.

Design reviews are focused on several factors. Some of these may include items shown in Table 8.6.

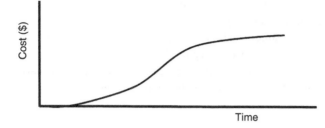

Figure 8.4 Costs Vs. Time. Figure developed by David Tennant

Table 8.6 Design Review Criteria – Sample

	Item	Description
1	Product objectives	• What are the product objectives approved by Senior Management? • Do the business case objectives match those approved? • Is the design meeting those objectives?
2	Features	• What features are required by stakeholders? • Are the technical requirements feasible?
3	Useability	• Will the design meet the intended market needs (geographic, income based, etc.)? • Will the product be aesthetically pleasing? • Is the product durable?
4	Ergonomics (human factors)	• Is the product comfortable to use? Will it fit the 90% population profile? • If using a display, are the colors coded per standards and industry guidelines? • Is the display concise and uncluttered? • Is the product easy to understand and use? • Is it possible to misuse the product (by accident or intentionally)?
5	Technical review	• Have appropriate safety factors been incorporated? • Has the product been stress modeled? • Has a prototype been developed and tested to confirm engineering models? • Are the materials used in the product appropriate and long-lasting? • Is it possible to misuse the product in a technical sense? • Have abuse protections been built-in?
6	Constraints	• Are we staying within the budget and schedule? • When will the design be "frozen?" • Have stakeholders had appropriate time for input?

Table developed by David V. Tennant

Modeling – Speeding Product Development

In the past, a new product would go through a long, extensive process to develop one or more prototypes. Designs points would be evaluated with calculations comparing a variety of materials to determine the best design. With this approach, how many prototypes failed? How many times were new designs – and more prototypes – developed at great expense and time?

Today, there are computer models that can evaluate or predict the performance and longevity of a product. For example, Figure 8.5 shows the expected temperature stratification in a boiler that produces steam for both the manufacturing process and to generate electricity. Earlier trials would have seen a boiler manufactured based on past history and new design concepts. The new design would have to go through trials to determine its true performance. Boiler builders would many times offer a new boiler concept to a customer at no cost in order to evaluate its performance under real plant operating conditions. After 6 to 12 months, enough data would have been collected to confirm whether the design was good enough. Further developments in boiler design would be incremental and based on data collected from beta or test run boilers.

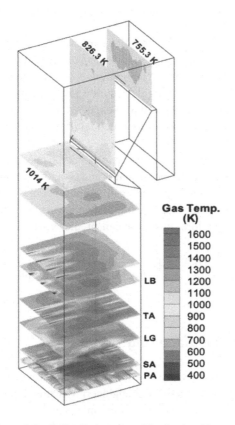

Figure 8.5 Boiler Temperature Distribution. Photo courtesy of Gerald Elliott, International Applied Engineering

It is known that previous designs are "over designed" for example using stronger, thicker materials than needed. This does not make the product necessarily safer, but definitely more costly. In other cases, some materials may have been misapplied, leading to product failures. The primary point is that modeling can help improve design and performance of products using fewer prototypes.

Advantages of Engineering Modeling

- Reduces the number of prototypes that must be constructed.
- Allows the designer to optimize materials and shapes.
- Can predict with a high degree of accuracy the material strength and failure points of the product.
- Can predict how the product will operate under a variety of stressful scenarios.
- Allows designer to test different materials or shapes to enhance or prevent heat transfer.
- Will decrease development time.

Figure 8.5 shows a model of a recovery boiler. Most paper pulp plants use a recovery boiler to dissolve organic residue left over from the combustion of wood. In the recovery boiler, heat produces high pressure steam, which is used to generate electricity in a turbine. The turbine exhaust, which is a lower pressure steam, is used for process heating.

Combustion of black liquor in a recovery boiler needs to be controlled carefully to avoid high concentrations of sulfur. This can lead to the production of sulfur dioxide. Properly controlled emissions will lead to efficient and environmentally cleaner combustion. The use of engineering models has been highly accurate in predicting not only the temperature distribution within the boiler but also expected air emissions. This is a typical application at facilities that manufacture cardboard boxes and various grades of paper.

Modeling is also appropriate for a variety of scenarios where a prototype may not be possible. For example, nuclear waste from nuclear power plants and weapons production can stay radioactive for thousands of years. By modeling different types of storage casks, an optimum design can be reached. It would be impossible to test a storage casks prototype for a thousand years; therefore, modeling can predict strength and integrity into the distant future.

There may also be scenarios where prototyping a product in a hazardous area may not be possible. Again, engineering modeling can assist in reaching the optimum design.

Integrating Supply Chain and Manufacturing

New product development may require new materials, equipment, software, suppliers, or consultants. It will be necessary for the R&D and engineering teams to work closely with the supply chain group to ensure the correct components and materials are procured. Note that some companies may still refer to supply chain as "Procurement."

What is the purpose of supply chain? Simply, Supply Chain is responsible for the procurement of products and services external to the company. Depending on the size of the business initiative, this can run into hundreds of millions of dollars. Therefore, supply chain may play an important role in new product development, which must be coordinated with the engineering team and other functional areas.

As stated earlier, publicly held companies must comply with Sarbanes-Oxley financial reporting requirements. This has a direct impact on how Supply Chain operates within a company. The following is a true case in which the author has direct knowledge.

Case 8.1 Reasonable Procurement Action or Big Mistake?

The development of the firm's new product, a specialty cable, was behind schedule with delays in both R&D and product testing (Quality Control). The new facility under construction to produce the new product was also behind schedule. Consequently, there was pressure from senior management to "get the job done."

It was determined early in the project that external engineering services would be required to assist with environmental design, permitting, site drainage, and air emissions. The engineering services had been approved in the project's budget and the Product/Project Plan. The following conversation took place between the Product Manager and the Project Manager regarding obtaining engineering services.

PROJECT MANAGER (PM):	Hello Bill, I have several requisitions here that need your approval.
PRODUCT MANAGER (BILL):	OK, let's see, we have five requisitions totaling about $300,000 dollars. Are these services in the budget?
PM:	Yes.
MILL:	Is the scope of work acceptable and within the project's scope? (As Bill begins to sign each order)
PM:	Yes.
BILL:	Great, here you are (handing back the signed orders). When will the work begin?
PM:	Well, the work has already been completed. I did not want to go through official channels as that would have taken more time and we're already behind schedule. I gave the engineering consultant a verbal order over the phone. They have a good history working with us, so they were content to accept a verbal. I just need these orders approved so when the supplier's invoices hit our system, we'll have the necessary back-up, and they can get paid.
BILL:	This is unacceptable. You have violated company policy and could get us in trouble. This will not be tolerated in the future. All POs are to be signed off by me before you begin work next time. If this happens again, I will relieve you of your position. Understood?

Note: This company is publicly held; that is, its stock is traded on the US stock exchange.

Questions for this case:

1. Was the project manager justified in issuing the purchase orders verbally to save time?
2. What potential repercussions can occur from this action?
3. Is this purchase acceptable since Bill has now approved the requisitions?
4. Since this item is in the approved budget, is this action acceptable?
5. If this had been taken through the procurement process, what activities might have occurred?
6. Is there a risk by the engineering company accepting an order verbally without any paperwork?
7. What is the possible liability of both the company and the engineering firm?
8. What are the potential repercussions for both the PM and/or Bill?
9. How would your company view this situation?

The Role of Supply Chain in Product Development

It is important to note that Product Development of any kind will likely need to purchase outside assistance, whether in services or products. The Supply Chain department generally has the authority and the expertise to help the Product Manager with obtaining the needed items in a timely manner. The Supply Chain department usually has a contracts group, with attorneys specialized in this area to generate the necessary contracts.

For many large companies, supply chain provides a valuable service with considerable expertise. Depending on the size of the new product or business initiative, the Product Manager may have a procurement specialist assigned exclusively to support the program. Some of the typical activities performed by the Supply Chain organization include:

- Maintenance of an "approved" supplier list
- Auditing of suppliers
- Issue of RFIs, NDAs, and RFPs
- Reviewing proposals
- Negotiating prices and statements of work
- Review and issue of contracts and change orders
- Expediting orders that are running late
- Working closely with technical personnel to understand project needs
- Directing and managing the business issues of procurement
- Ensuring contracts are closed at conclusion of supplier's work

Each of these is discussed in further detail below.

Maintenance of an "Approved" Supplier List

It is fair to assume that most companies have awarded work to suppliers who were either not qualified or lost the capability to provide key services. Many times, it is late in the schedule when a supplier must be fired and/or replaced. At this point, significant time and dollars have been expended and are not recoverable.

One way to avoid supplier problems is to have a database of approved suppliers. This is a common activity for many companies. If a supplier wishes to be placed on the approved list, they must go through a review process. This may include (but is not limited to):

- An audit of their finances – it is not desirable to place companies on an approved list if they are approaching bankruptcy or facing liquidity issues.
- A tour and review of their manufacturing facilities to determine:
  - Is the facility well-run?
  - Is housekeeping neat and orderly (related to safety)?
  - What is the safety record of the facility?
  - Does it have adequate manufacturing capacity?
  - Is there a preventive maintenance program?
  - How robust is their QA and QC program?
- A review of references: i.e., follow up with their clients to determine their satisfaction
- Do they have service centers or a dealer network nearby?

Issue of RFIs, NDAs, and RFPs

Suppose you require a very specialized expertise, say a company that does FEA modeling (Finite Element Analysis) for the new product you are developing. If a company with this type of expertise is not on the approved supplier list, how do we find one? Your procurement specialist may do a search of companies (with technical input from you) that offer this service and then issue a Request for Information (RFI) to each one. This essentially gives the suppliers the opportunity to describe their expertise in this area. Based on their responses, this will allow the Product Team to determine which companies should receive an RFP.

If it is determined that these companies offer a valued service, it may be useful, but not required, to place them on the approved supplier list.

A Non-Disclosure Agreement (NDA) is used when your company desires to share confidential or proprietary information with a suppler (who needs the information to perform their work). An NDA is a signed acknowledgement that each company will keep key information confidential. This is a legally binding document (for both sides) and may even have a timeframe specified.

An NDA usually addresses protections for customer databases, proprietary and intellectual property, types of business or manufacturing processes in use or under development, and company strategies or future plans. An NDA may also have an expiration date. For example: "This NDA will be binding on both parties for three years from the date of signature."

As one can expect, an NDA is extremely important for a company that is developing a new product. Should a firm's proprietary information be released – whether intentionally or by mistake – the consequences can be very damaging.

Companies that have passed a review and are considered to perform the work will receive a Request for Proposal (RFP). The RFP is a formal document that describes the following:

- Work to be performed
- The timeframe
- Location
- Proposed contract
- Terms and Conditions
- Technical and commercial points of contact
- Technical specifications
- Standard proposal form (to be filled in by bidder)
- Instructions to bidders
- Deadline date the proposal must be returned to buyer.

Specifications

Specifications are detailed and graphic information that describes, defines, and "specifies" the services or items that are to be provided. Specifications are generally technical and very detailed. There are several types of technical specifications that can be developed. These include:

- Design Specifications
- Performance Specifications
- Functional Specifications.

Design specifications are developed by the buyer and will describe the physical attributes of the deliverable. A description of its operating environment may also be included.

> As an example, suppose you have a chemical testing lab and need a specialized pump that will operate in a hazardous location. Your firm's engineering department has developed a pump design and has tasked a local pump manufacturer with building the pump. Your design team has specified the operating conditions, the size and material of the pump casing, size of pump inlet and outlet, the size and material of the impeller, motor size, hazardous electrical circuit box, wiring diagram and connections.
>
> Upon installation, it is apparent that the pump is not working. Whose fault is this? First, if the pump manufacturer built the pump to your firm's design specifications, the fault will lie with your firm (buyer). Be aware that this type of specification scenario puts the risk on the buyer (you).

Performance Specifications are developed by you (the buyer) but the manufacturer (the supplier) is given the freedom and responsibility for designing, building, and delivering the product.

> Let's return the previous pump example. However, instead of developing a design specification, we will provide the supplier with a performance specification. In this case, the buyer will leave all design and material decisions to the supplier.

For this example, the supplier will simply provide the operating characteristics:

Table 8.7 Sample Functional Specification No. 1.

Operating flow and temperature	• 40 gallons per minute (gpm) • 60^0 F to 80^0 F
Fluid characteristics	• Liquid with a dynamic viscosity of 1.12 to 0.80 mPas • Ph content of 2.5 to 5.5 (very acidic)
Operating pressure	• Maximum pressure = 30 psig
Operating environment	• Clean lab with negative pressure • Potential low concentrations of chlorine gas • Gases in air may provide an explosive environment
Operating time	• Continuous for up to 40 hours.
Other requirements	• Explosion-proof wiring connections and junction box • Emergency pump shut-off circuit • 120 v, 60 Hz. power available • Inlet and outlet pump isolation valves

Table developed by David V. Tennant

This means the supplier will design, build, and provide a pump that meets the supplier's performance requirements. Note that this shifts the risks away from you (the buyer) to the supplier. Table 8.7 illustrates a sample functional specification.

Functional specifications can be considered a subset of performance specifications. This type of specification is more commonly found in large enterprise-wide computer IT applications, plant control systems, robotics, and data acquisition systems. A functional spec describes the function and features of the deliverable. The design and performance risks are on the supplier.

Table 8.8 is a sample functional specification we might see with a manufacturing plant control system. Note: an actual functional spec would be very detailed and most likely run up to 100 pages of requirements including an instrumentation list; this is simply a small sample.

It is appropriate to note that the technical issues are usually discussed directly between the Buyer's engineering group and the Supplier's technical team. The business issues are generally handled directly between the Supply Chain point of contact and the Supplier's sales or contract group.

However, the buyer's engineering and supply chain groups must work together to ensure the appropriate equipment or services are procured that will fill the company's requirements. This requires extensive communications between these two groups. For example, it is likely that the engineering team will write the technical specification and statement of work. The supply chain group will provide business terms and conditions, draft contract, and other business documents. These will be combined into an RFP (request for proposal) which Supply Chain will issue to one or more suppliers.

Table 8.8 Sample Functional Specification No. 2.

No. of operator interface screens, 40"	12
System inputs	150
System outputs	150
Field instruments	Transmitters, controllers, positioners, etc. Control valves Motor controllers
Deliverable	A fully functional manufacturing control systems that will control start-up, normal continuous operations (24/7), and shutdown (normal and emergency conditions) under a variety of operating scenarios.
	The system will comply with industry standards for operations, human-factors (ergonomics), and safety.
	For critical plant operations, two out of three signal logic will be provided.
	The system will have an "automatic" operating feature but can be overridden or switched to "manual" operation, when needed.
	The control system supplier will develop all control logic, algorithms, and computer code to safely operate the plant with high reliability and safety.

Table developed by David Tennant

Proposals, Pricing, Statements of Work (SOWs)

All proposals submitted by potential suppliers are returned to the supply chain group. It will be a joint effort by Supply Chain and Engineering to evaluate the proposals from both a technical and business viewpoint.

Technical considerations:

- Delivery dates
- Compliance with technical requirements and operation
- Performance of equipment (or qualification of consultants if services)
- Product technical support
- Will it meet the application we are seeking?
- Ease of operation and maintenance
- Track record of the equipment or system.

Business considerations:

- Pricing
- Terms and Conditions (pricing tied to schedule milestones?)
- Contractual issues
- Change orders
- Termination for cause or convenience
- Statement of Work
- Past history, if applicable, working with this supplier.

Supply Chain will negotiate final pricing and provide contracts for signing. Once both parties sign, work can begin. However, Supply Chain does not simply back out when the contract is in place. It is likely that Supply chain will be engaged throughout the life of the project/product's development to assist with change orders, compliance of supplier with the contract, and closure of supplier's work orders.

Change Orders, Expediting, and Contract Closure

It is highly likely that there will be changes in the project's scope during the life of the product's development. It is the author's opinion that there will always be changes in the scope of a project. If processes are set up to manage changes, then it is unlikely that scope creep will get out of control. And one should not assume that changes in scope always increase time and costs. There are instances where a small change (and incremental cost) may in fact reduce the product's development time and consequently cost less. An example of this might be paying an extra $50,000 for FEA (finite element analysis). This could easily reduce schedule time, assist in developing cost-effective warranty policies, and save money by reducing the need for prototypes.

Figure 8.6, Change Request Review Process, is a simple process to evaluate a requested change order.

Expediting is when supply chain gets involved to work with suppliers who anticipate late delivery of the products or equipment. The engineering and R&D groups are busy developing and designing the new product and have little time to deal with delivery issues. Supply chain, however, has a role is holding suppliers accountable. And being familiar with the contract allows the expediter to apply leverage. Delivery slips may be beyond the control of the supplier, so it may be necessary to reach an agreement of new, firm delivery dates. However, the expediter will generally do everything he or she can to hold suppliers to the terms of their contract.

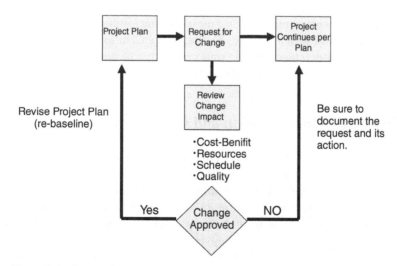

Figure 8.6 Change Request Review Process. Figure developed by David Tennant

Figure 8.6 Change Request Process, is a simple process to evaluate a requested change order. A change order can be suggested by anyone on the project (supplier team or buyer team), but each change request should go through a formal review process.

This will allow the Product Manager to control the scope of the project more effectively. While change orders can be technically or commercially oriented, it is the Supply Chain group who issues the C.O. (Change Order) to the supplier and helps negotiate pricing. This is typically an ongoing activity.

Once a product has completed its development (or the project is over), there is significant paperwork that must be completed. It is customary to notify the supplier that the deliverable has been accepted and the contract will be closed by a certain date. This is also the signal to the supplier to submit any final invoices for payment. Supply Chain, Engineering, and Accounts Payable should work together to confirm final pricing is appropriate and within the contractual boundary.

New Products Vs. Commodities

Recall from the product life cycle, that competitors will eventually catch up and offer serious competition to your once-dominant product. What exactly is a commodity? In the traditional sense, a commodity is considered a raw material that is grown, mined, or produced in mass quantities. How does a company maintain its dominant position?

When a new product is introduced to the marketplace, it will enjoy a dominant position for some time. If we consider new cars, they attempt to maintain their dominance by advancing the technology (think Apple or Android in your car), styling, features, and performance every three to four years. While there are many competitors in this space, those that are deemed to have higher quality, or the perception of higher quality, will tend to have market dominance OR command higher prices.

To a consumer, a limited number of choices means he or she will pay a higher price. When a product becomes a commodity, the consumer knows there are many similar products and the deciding factor will then be price. This is where market segmentation (discussed in Chapter 2) comes into play. This means adding features that appeal to a certain demographic group such as age, gender, location, etc., or to offer with added value. People will pay more if they perceive your product has higher quality, a better warranty, better features, or is a recognized brand.

New Technologies – Identification and Adaptation

A number of futurists have indicated that technology changes in the next 10 years will have a greater impact on society than the previous 100 years. Indeed, if one considers the evolution of technology, the impact in just than last 25 years has exploded. Our smart phones now contain more computing power than the Apollo spacecraft that landed on the moon in 1969. Our phones allow us to not only communicate, but take pictures, surf the net, shop, track our appointments, and dozens of other "apps" that can be downloaded to make everyday living more convenient and entertaining.

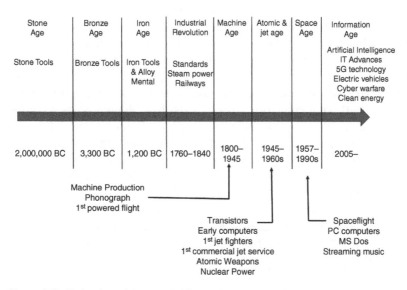

Stone Age	Bronze Age	Iron Age	Industrial Revolution	Machine Age	Atomic & jet age	Space Age	Information Age
Stone Tools	Bronze Tools	Iron Tools & Alloy Mental	Standards Steam power Railways				Artificial Intelligence IT Advances 5G technology Electric vehicles Cyber warfare Clean energy
2,000,000 BC	3,300 BC	1,200 BC	1760–1840	1800– 1945	1945– 1960s	1957– 1990s	2005–

Machine Production
Phonograph
1st powered flight

Transistors
Early computers
1st jet fighters
1st commercial jet service
Atomic Weapons
Nuclear Power

Spaceflight
PC computers
MS Dos
Streaming music

Figure 8.7   Technology Advances in History. Diagram developed by David Tennant

Consider Figure 8.7 which charts the evolution of technology in civilization.

Note that very little progress was made in the first two million years of human existence. Yet advances in technology over the last 300 years have increased at an incredibly fast rate.

Little wonder that as time marches forward, the advances in technical achievement are accelerating. As a firm concerned with product development, how do we anticipate how to incorporate or enhance these advances? What will customers expect?

There are a number of ways to research and become aware of new advances. Obviously, searching the internet for new technology trends has made it easier to find new tools, advances, and breakthroughs.

Almost every industry has trade groups that sponsor conferences, seminars, training, and other means to stay abreast of new applications. These new discoveries, or their applications, are often published in trade journals, magazines, etc.

Universities perform a significant amount of research which is funded by private corporations, public corporations, and government agencies. Many times, if not proprietary, their results are published in scientific journals.

Finally, there are a number of futurists that have an uncanny ability to anticipate advances in trends or technology which can provide companies data on where to concentrate their efforts in new products and services.

Here are a few of advances that are still in the development stages but have huge potential for the future. How can entrepreneurs participate?

Electric Vehicles and Smart Cars

Currently, as of this writing, EVs comprise only 2% of the US market for new cars. As a result of government actions to mitigate climate concerns, a number of governments have mandated that all new cars must use a clean form of energy, such as electric batteries, hydrogen, natural gas or similar. In other words, the internal combustion engine – which has been in use for over 100 years – is on the way out. Think back to when travel by horse buggy was displaced by the automobile. We are now seeing a similar shift to newer and cleaner technologies. Thus far, the following countries are on track to eliminate conventional gas- and diesel-powered cars within the next 20 years: Germany, Britain, India, Norway, and France. Other countries, while not outlawing combustion-engine cars, have definite sales goals for alternative-fueled vehicles. These include Austria, China, Denmark, Ireland, Japan, the Netherlands, Portugal, Korea, and Spain. While the United States does not have a mandate for EVs, there are tax incentives in place for their purchase and most major auto makers are moving in the direction of EVs. This is being encouraged through public policy.

This means there will be plenty of opportunity for suppliers to develop new products and services to support this newly developing market. Imagine how many new fast-charging stations will be required across the USA and other countries to support a growing EV market.

5G Networking (5th Generation)

With the widespread use of smart phones, 5G networks will allow increased internet speeds and low latency. As with any new technology, there are advantages and disadvantages. Some of the advantages include high resolution and large bandwidth, support of huge data loads and streaming (gigabit size), consistent networking worldwide, and will be an integral technology to allow the use of driverless cars and promoting smart cities (see below).

Some of the disadvantages include high infrastructure costs. The 5G towers will need to be positioned closer together meaning many more cell towers (at great cost). It is likely that current cell phones may not be 5G compatible, meaning that some people will need to purchase new, expensive 5G phones. The implementation of 5G networks is occurring in urban and suburban areas where the costs may be spread over many potential customers. However, where does this leave rural areas?

The technology can assist drivers avoid accidents. For example, if cars can use GPS positioning with 5G, it is feasible that they can "sense" where other cars are located nearby and avoid collisions.

Issues that still need to be addressed for 5G include personal data privacy and cyber security.

Smart Cities and IoT (Internet of Things)

A smart city is connected with cloud-based applications to better run a city's infrastructure. This can include better city services, reduced traffic congestion, efficient energy distribution, and pairing the town's residents through cell phones. This will very likely be the case when cars are connected and able to be routed in the most efficient routing. For example, the IoT can connect and direct traffic lights to help traffic flow and also guide your car to the nearest parking spot close to your destination. Your smart phone will also contain your driver's license. Currently, a number of phone apps can plot and direct you to a destination in the fastest route.

Pair this new infrastructure with smart grid technology and the options are limitless. A smart grid is a planned local or national network that uses information technology to deliver electricity efficiently and with high reliability. This new network will allow the two-way flow of energy and information. There are also communities that are connecting using a micro-grid. A micro-grid is a small, localized electric network that can disconnect from the wider state or national grid and operate on its own. One advantage to this is that the micro-grid can operate with its own power source, connecting local homes and businesses when there is a power outage or storm in an adjacent area.

The above technologies are just a few of the major "disruptive" systems that are gaining traction on the federal, state and local levels. Not only are governments investing in these technologies, but companies as well. Note that this is happening worldwide, not just in Europe and North America.

For companies that have creative and visionary leadership, it is an exciting time to be in the business of new product development.

Alignment with Business Strategy

Strategy is concerned with the future and changes in capabilities, products, or services to meet anticipated market conditions. In Chapter 2, the importance of corporate strategic planning was discussed. As such, all projects, including new product development, should support the business strategy. Think of this also as the Corporation's Strategy for future profitability.

Do new product development projects get canceled? The answer is "yes," all the time. What can lead to cancellation? There are many factors that can lead to project cancellations. These include:

- A change in company leadership. This is more common than some people might think. When leadership at the top of a company changes, usually, that means change is coming. New leadership is brought in when the previous leadership is not meeting results – usually financial results. As a result, the new leaders will evaluate the company from top to bottom and chart a new direction. This will depart from the old paradigms and culture. Many of the current projects, including new product development will be re-evaluated; and it is likely that some of these projects will be shut down. On the other hand, it is also possible that other new products – that meet the new direction – will be initiated.
- Significant budget and schedule overruns. Recall that all products must recover their investment and make a return when launched in the marketplace. The marketing group will typically determine the appropriate margin, competitive landscape and pricing for the new product. Let's consider both budget and schedule issues:
  - **Budget**. If the development of new products or services goes significantly over budget (say greater than 50%), it is likely that the costs of development along with a reasonable profit margin will never occur. In this case why would the company continue to support its development? One may ask, why not simply sell at a price to recover a ROI (return on investment)? The answer is simply because your competitors may be significantly cheaper, and you will never make enough sales

margin. Or, even if you do not have any current competitors for your new product, you will find competitors will eventually emerge.

The costs of new product development will be closely tracked by the finance department (mostly likely the CFOs direct reports). The key yardsticks that will be used to measure product feasibility are NPV (net present value), break-even point, ROI (Return on Investment) and margins. All of these are significant measures that must be considered together.

Note that different products will have a different timeframes to reach profitability. Some industries may desire to see a payback for newly launched products within three years. Others may find a much longer timeframe acceptable.

- **Schedule**. Budget and schedule are closely related as in the axiom "time is money." If capital funding is being borrowed to develop the new product (for example, adding manufacturing capacity, or using new testing equipment or software), then the longer a project drags on, the more costs will be incurred.

  There is also the possibility that competitors will be first to market; thereby scoring marketing and PR visibility. Remember, first to market usually gets market share. Therefore, it is important that your new product should meet or beat its scheduled market release.

- Downturn in the economy. The economy tends to be cyclable with upturns and downturns happening over long time periods. Further, unpredictable events, such as the Covid-19 pandemic of 2020–2021, are not foreseeable but can have a devastating effect on company finances. When company revenues (and therefore profits) are strained, it is likely that some projects will be put on hold or canceled. This type of event is beyond the control of the product team, even if on budget and schedule.

Using SWOT

It is very common for businesses to perform a SWOT analysis in developing new products or even in strategic planning. SWOT stands for Strengths, Weaknesses, Opportunities, and Threats. This is a common business tool that can be as simple or as sophisticated as needed. The author has seen both types in practice with the more sophisticated SWOTs utilizing spreadsheet data and supporting documentation.

A SWOT analysis can be used in many different scenarios:

- Considering entrance into a new market
- Launching a new product or service
- Surveying your firm's role in current market
- Evaluating your firm's current position in a strategic plan (how should we change in getting from point A to point B?)
- Comparing your firm to your competitors (competitor analysis)
- At the beginning of a project to determine relevance
- Potential merger and acquisition activity.

If we consider a simplistic approach, the following will serve as an example.

Let's assume that we are developing a new product – Product Zed. As part of our initial review, the product team has performed a simple SWOT as shown in Table 8.9.

Table 8.9 SWOT for Product Zed.

Strengths	Weaknesses
• Highly experienced product team • Financial and Sr. Management support • Company has experience in the market • Company reputation is strong for this type of product • Company is financially strong • Company holds a leadership role in this space • New efficiencies and improvement to manufacturing plant will lead to lower costs	• Some personnel turnover on team, will need to find experienced replacements quickly • Market for this product may be maturing • Feeling pricing pressure from our customer base • Current model is showing its age (market maturity) • Our sales team will need to undergo extensive training for new, replacement model
Opportunities	Threats
• Still have market leadership which can be exploited with excellent marketing • Creative team has great ideas to make a tremendous leap in product successor • Competitors still not a large factor in market and do not understand our customer base	• Foreign competition is making minor inroads to our market • Manufacturing costs of competitors is cheaper • Competitors have worldwide marketing expertise

Table developed by David Tennant

Gates and Stakeholders

As noted earlier, a stakeholder is a person who can have a positive or negative impact on your project or be impacted by your project. Identifying all possible stakeholders is desirable, but some will be more important than others. Those stakeholders that are at a high level or have a predominant role in your product's development will wield a great amount of influence. This can include project sponsors, company executives, and JV (joint venture) partners.

These key stakeholders will need to be kept informed of the product's development, status of budget and schedule, and testing results (QA/QC). This can be accomplished by using gates as review points. Recall from Figure 8.8 that "gate" review points are shown in the project life cycle. These are go or no-go decision points where the executive team reviews the status of the project or product's development; a decision is made whether to continue forward or not.

While there are only five gates shown in Figure 8.8 Potential Project Gate Reviews, it is likely that a real product development project would have many more than this. At each gate, an evaluation should be performed to determine if the project is still worth supporting, and enough funding is provided to reach the next gate.

At each gate, the following items should be evaluated:

• Budget status
• Schedule status
• Design status
• Latest risk review results
• Justification why product should still be funded
• Does project still fit with company strategy?

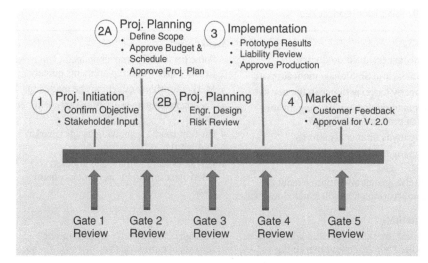

Figure 8.8 Gate Reviews. Figure developed by David Tennant

- Are there any changes to product's intended purpose or features? If so, why?
- Has SWOT been updated to account for new business conditions?

Typically, when the product team passes through a gate successfully, funding to get to the next gate is provided. However, each company has its own rules for gate control.

Chapter Key Points

- Developing a new product is a project.
- All projects follow a lifecycle.
- Performing risk reviews is an important activity to identifying potential problems.
- Design reviews should occur at well-defined points during product development.
- Changes in design become more costly as you go further into development.
- Engineering modeling can be used to shorten development time and it should reduce (not replace) the number of prototypes that are required for testing.
- Supply chain (procurement) will play a key role when purchasing large-dollar capital equipment.
- Specifications are generally developed by the engineering group and are necessary in providing a technical description of a product's operation or design.
- Budgets and schedules should be closely tracked to ensure costs do not jeopardize your product's continued support.
- A SWOT analysis can assist in positioning a company in launching a new product or service, and in comparison, with competitors and market trends.
- Stakeholders should be informed of a products status during development and launch.
- Gate reviews can provide clarity in determining a project's status and informing stakeholders.

Chapter Discussion Questions

1. Why would we consider the development of a product a project?
2. What are some of the activities that occur in planning a project?
3. Risk reviews can be quantitative or qualitative. Under what conditions would we use a qualitative review? A quantitative review?
4. What is the purpose of a risk review?
5. What is the purpose of a design review?
6. Why is a contract important in dealing with external suppliers?
7. Under what circumstances would we see a new product's development canceled?
8. Why is it important to track budgets and schedules?
9. What is a SWOT and how often should one be performed?
10. What is the purpose of a Gate review?

Answers to Case 8.1 Reasonable Procurement Action or Big Mistake?

1. Was the project manager justified in issuing the purchase orders verbally to save time?

   The project manager was not justified in his actions. He violated company policies, and in this case, also violated Federal Law: Sarbanes-Oxley requirements.

2. What potential repercussions can occur from this action?

   There are a number of repercussions that can occur:
   - *The SEC (Securities and Exchange Commission) could fine the company*
   - *People could lose their jobs over this action*
   - *The company could refuse to pay the invoices*
   - *The supplier could take the company to court if invoices refused.*

3. Is this purchase acceptable since Bill has now approved the requisitions?

   No, this purchase did not go through appropriate purchasing channels. It is Bill's job (as the Product Manager) to adequately supervise his team, be aware of their actions, and be aware of company policies. It is also his role to ensure that his team is aware of company policies and procedures.

4. Since this item is in the approved budget, is this procurement now acceptable?

   No. It is never acceptable to violate or go around company policies. This is especially true for this huge dollar amount ($300,000). Company policies are in place for a reason.

5. If this had been taken through the procurement process, what activities might have occurred?
   - *The services may have gone out for bid in order to obtain competitive pricing. However, with justification, a sole-source contract can generally be approved to accelerate an award. Schedule sensitivity can be a legitimate justification.*
   - *At the least, a contract would have been in place outlining the responsibilities of each company (buyer and supplier).*
   - *Even if a sole source contract, it is likely that price negotiations would have occurred, and penalties would be in place for non-performance.*

6. Is there a risk by the engineering company accepting an order verbally without any paperwork?

There are a number of risks in accepting a verbal order. The buying firm may dispute costs, scope of work, schedule, final quality, etc. If there is nothing agreed to in writing, this represents a huge risk to the supplier. If the buyer company changes its mind on scope, timing, features, etc., how would these changes be negotiated and paid for? A verbal order may encourage scope creep by the buyer and the engineering firm has no legal or business remedy. Note: scope creep (i.e., failure to control project scope) is a major factor in project failure.

7. What is the possible liability of both the company and the engineering firm?

- *There may be disagreements over the quality or scope of work. This may lead to legal action by either side.*
- *What if the design does not work? Without any metrics or testing in place, it will be hard to prove who is at fault and may lead to litigation.*
- *Without a contract, there are no legal protections for either party.*

8. What are the potential repercussions for both the PM and/or Bill?

The company may fire, or demote, one or both of them. This is not a stellar career moment.

9. How would your company view this situation?

The author suggest that most companies would take a very dim view of this scenario.

Answers to Chapter Discussion Questions

1. Why would we consider the development of a product a project?

The development of a product is a project as it follows the project lifecycle. Further, all product development projects require a budget, schedule, resources and should be tracked and measured. If the development of a product is not managed as a project, it is likely to be a chaotic situation with a high risk of failure.

2. What are some of the activities that occur in planning a project?

- *Developing and confirming objectives and metrics*
- *Producing budgets and schedules*
- *Assigning resources*
- *Managing supply chain activities*
- *Performing risk reviews*
- *Shepherding projects through gate/status reviews*
- *Communicating with stakeholders and suppliers*
- *Ensuring QA and QC of the final deliverable.*

3. Risk reviews can be quantitative or qualitative. Under what conditions would we use a qualitative review? A quantitative review?

A qualitative review is quicker and easier to perform. It is also easier to communicate results. This should be used when the confidence level of cost impacts is unknown.

A quantitative review would be appropriate when a high level of confidence in the cost impacts of risks can be determined. It is also useful in developing a contingency budget. Finally, companies that do work for the federal government are usually required to perform quantitative risk analyses.

4. What is the purpose of a risk review?

The purpose of a risk review is to identify potential future problems. If we can identify problems in advance, it may be possible to develop strategies to avoid or minimize them. This will reduce the "firefighting" that develops in so many projects.

5. What is the purpose of a design review?

A design review occurs at several points during the product's development. It is similar to a gate review but is focused primarily on the technical and engineering aspects. For example, it may be determined that a formal design review will occur at 25%, 50%, 75% and 90% of a product's engineering development. It is likely that the engineering team will have weekly meetings, but a formal design review usually involves senior management, clients, and other key stakeholders – including the possibility of suppliers, when needed.

The design review will confirm that the features and specifications are still relevant. It also provides the opportunity for the company to request changes (at least early in the project) such as materials, colors, texture, function, etc. It is the design team's role to point out any quality issues, whether the design and manufacturability are achievable, and to confirm that original objectives and design basis are still in effect.

The design review also gives the engineering team the opportunity to hear feedback from marketing and sales groups about changing customer preferences, if any. It is an excellent opportunity to communicate across the organization what the ultimate product will look like, how it will perform and to identify potential issues with the product or it's development.

6. Why is a contract important in dealing with external suppliers?

A contract describes what a supplier will provide for a given price and when. It obligates the supplier to provide a product or service within key technical specifications. It also obligates the buyer to pay for the product or service upon delivery.

A contract protects both the supplier and buyer from misunderstandings as it clearly spells out the expectations and obligations of each party.

7. Under what circumstances would we see a new product's development canceled?

- *When new leadership is appointed to the company.*
- *When the costs of development outweigh any chance of a return on investment.*
- *A downturn in the economy may reduce a company's ability to fund new product development.*
- *A competitor beats you to market with a better, cheaper product.*

8. Why is it important to track budgets and schedules?

If the budget and schedule are out of control, there is a possibility your new product will never be profitable. Why would a company continue to support development of a product that will be a money loser?

9. What is a SWOT and how often should one be performed?

SWOT is an acronym for Strengths – Weaknesses – Opportunities – Threats. A SWOT review should be performed on a regular basis (at least yearly) for companies that practice strategic planning. It is also helpful to perform a SWOT review prior to developing a new product, business initiative or joint venture partnership. A SWOT helps a company determine its position in the marketplace relative to its competition.

Finally, it is very common to see a SWOT review in a business case for the reasons above. A business case is a feasibility study that presents several options for consideration. It will assist management in making an informed decision.

10. What is the purpose of a Gate review?

All projects should go through a gate review process for the following reasons:

- *To ensure that the new product is still relevant to the company's strategic business plan*
- *To get a gauge on the product's status (is it on budget, schedule, etc.)*
- *To hold the product team accountable*
- *As a justification for continued support and funding by senior management*
- *Typically, there are multiple gate reviews during the development cycle. It serves as a reasonable sanity check and can also communicate status to key stakeholders.*

Bibliography

Bangs, D. (2005). *Business Plans Made Easy*. Entrepreneur Media.

Hill, G. (2010). *The Complete Project Management Methodology and Toolkit*. Boca Raton, FL: CRC Press.

Levi, D. (2007). *Group Dynamics for Teams*, 2nd e. Thousand Oaks, CA: Sage Publications.

Norman, D. (2013). *The Design of Everyday Things*. New York, NY: Basic Books.

The Project Management Body of Knowledge, 6th e (2017). Project Management Institute, Newtown Square, PA: The True, S. (ed. 2011) *Business Acumen*. Kennesaw, GA: Coles College of Business.

9

Successful Product Launch and Post Review

The launching of a new or revised product involves elements of costs, expected demand, customer feedback, advertising, and marketing. While TV and radio ads, and print brochures all have their place, marketing today is much more sophisticated and involves the use of multiple channels, targeting, and social media.

As much thought needs to be concentrated on reaching your target markets as was done during the development of your new product. The best product on the planet will go nowhere without a market made aware of its existence. Indeed, the corporate landscape is littered with failed products that were released by smart people working at successful companies. Consider the following:

- Sony Betamax (1975) was released as "the" system for people to record their favorite TV shows, movies, etc. However, all of Sony's competitors were issuing recording systems using VHS tape which was more convenient and had a minimum two-hour run time. Sony had to abandon their system as the industry standard became VHS. Of course, VHS has been supplanted by digital recording on chips. This represented a failure to recognize a market trend.
- It was a good strategy for Harley Davidson (HD) motorcycles to issue clothing and personal accessories as an extension of the brand. This included tee shirts, bandanas, key chains, and other fashion items. However, most fans of the brand drew the line at Harley Davidson Cologne (1990). The failure of this product stopped HD from straying too far from their core market.
- The HP Touchpad (2011) was meant to compete with Apple's iPad®. However, the product did not offer anything new or different from Apple or other competitors and was only on the market for 49 days. This was a failure to offer a unique or better product.
- New Coke had won most of the taste tests that were conducted across the US in 1985. Consequently, Coca-Cola invested $4 million in the new product, but saw serious backlash from its loyal customer base. Within three months of the product's launch, the original formula Coke was brought back as "Classic" Coke. Regardless of the taste test results, Coke had failed to recognize the loyalty of customers to its flagship product.

Product Development: An Engineer's Guide to Business Considerations, Real-World Product Testing, and Launch, First Edition. David V. Tennant.
© 2022 John Wiley & Sons, Inc. Published 2022 by John Wiley & Sons, Inc.

These are just few examples of failed product launches. One can search the internet to find many more examples including Colgate launching a line of lasagna food and the Bic pen company getting into a line of underwear. Not entirely complimentary products.

Previously in this book, the topic of focus groups and prototypes were discussed. To launch a product, it would be desirable to have it tested by a target group to evaluate its potential acceptance. This would not be a test involving thousands, but rather a small subset of the market; perhaps 10 to 20 people. Customer feedback can be invaluable *prior* to launch. Some of the questions to be answered during customer testing include:

- What did you like about our new product?
- Was it easy to use?
- Were the controls intuitive?
- What would you consider a reasonable price for this item?
- Is this something you would tell your friends about?
- Would you buy more than one?
- Did you encounter any difficulties or negative experiences with this product?
- Do you use a similar competing product? How does this one compare?

These are a sample of questions that can be asked of the trial group. Based on responses, revisions to the original design or manufacturing may be necessary. Of course, this also depends on the complexity of the product.

It is important that final customer feedback is addressed, changes made (if needed) and the product has passed extensive testing prior to launch. Moving beyond testing, and prior to launch, what will the price of our product be? How much is too much?

What do we mean by profit?

- Gross profit: This is the profit a company makes after deducting the costs of making, transporting, and selling its products.
- Operating profit: The total earnings for a given period, excluding the deduction of interest and taxes.
- Pre-tax profit: The profit one makes by selling a product before taxes are deducted.
- Net profit: Sometimes called profit margin, or net margin, measures how much net income or profit is generated as a percentage of revenue.

For this book, when a profit or margin is discussed the reader should assume gross profit.

Pricing

There are many considerations in determining price. Among them, what is the company's objectives for this product? There are several different scenarios to think through as one contemplates a retail "price" for a product.

Competitor Positioning

This strategy presumes the product will be priced similar to competitors. This means you are benchmarking your firm's product against others. As an example, the gas industry, which is highly competitive and selling a commodity product, has pricing targeted to meeting or slightly beating competitors pricing. After all, natural gas is the same product regardless of where (or from whom) purchased. To survive in this market, not only must your "product" be competitive, but additional or superior customer service must be provided. Some gas providers offer a variety of rate plans to fit what their customers want. Another example includes the airlines. When one airline discounts (or increases) its fares, the competing airlines usually follow along by price matching.

Sales Positioning

This strategy believes that increasing sales will help in the long term and the firm is willing so reduce product profits. This has several perspectives. It may be that taking less profit with a lower sales price will discourage competitors from entering the market. It may also drive some from the market. Further, this may simply be an entrance strategy to entice customers with a low price and to later raise prices as the product takes root and gains market acceptance.

It must always be remembered that customers are very cognizant of price. They must perceive a *value* for the product in return for their purchase.

Figure 9.1 illustrates a typical demand vs. price graph. This shows that lower price will generally lead to higher demand. This does not show profitability: that is, we don't know the profit margin at price point P1 or P2.

The marketing group would most likely produce spreadsheet models based on price sensitivity, demand, and profitability.

It is entirely possible in Figure 9.1 that price 1 (P1) realizes more revenue due to higher sales but less profit per unit. If we think about grocery stores, their sales strategy is high volume with average margins of 2.2%.[1] Grocery stores can survive on such a small margin due to the huge volume of sales. Some stores will see revenue of $1 million per week. According to Yahoo Finance,[2] Kroger made $132.3 billion in revenue for the past 12 months (TTM – trailing twelve months) with a profit margin of 1.14%. This is for all 2,732 stores.

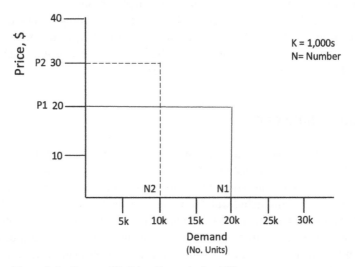

Figure 9.1 Demand Vs. Price. Figure by David Tennant

With a profit of 1.14% on revenues of $132.3 billion, Kroger made $1.513 billion in net income. Not bad for a low margin business.

Customer Positioning

This pricing model is focused on customer value, or perceived value. For example, a number of luxury car brands, among them Lincoln, Cadillac, BMW, Mercedes, Volvo, and others, wish to produce a limited number of units and charge a higher price. Although these models will serve the same purpose as a lower priced car – getting you from point A to point B – these brands are selling luxury and exclusivity. They are status cars and offer a more enthralling driving experience; and many times, are innovators in new vehicle technology. Remember that people will pay more for quality. Going beyond status are high performance sports cars, Ferrari, Porsche, Aston Martin, and others. These are very limited, high-performance cars with extremely high price tags (anywhere from $100 thousand to several million dollars). These will appeal to high-wealth performance car enthusiasts, collectors, and speculators (certain car brands will increase in value over time due to low production numbers).

It should be apparent that the car market is highly segmented, and pricing can be positioned to match those customer segments.

Profit Positioning

In this scenario, the company has predetermined what the return (profit) should be. Some companies will express the requirement for a 15-, 20-, or 25-percent margin. How is this determined? The company will determine the costs of the product and add an appropriate margin on top of this cost. For example, if our new writing pen cost $7.90 each to produce, then a 25% margin would be $1.98. Therefore, the price set for this product would be $9.88.

If we found that doubling the manufacturing quantity could reduce the production costs by 30%, we could similarly reduce the retail price (hoping for more sales) or leave the retail price as originally planned and obtain a higher profit.

However, a word of caution is needed. The simple act of adding on a percentage for profit may or may not be realistic. It depends on the product, your competitors, and market perception of your product. Remember, the market will ultimately set the price. All that you can control as a producer are costs.

Integrated Marketing

In the distant past, traditional marketing consisted or TV and radio spots, press releases, direct sales (relationships), brochures, and a company website. Today, in addition to traditional marketing, there are significantly more channels to reach potential customers. This includes all forms of social media such as Twitter, Facebook, LinkedIn, Pinterest, Reddit, and mailing platforms such as Mailchimp. Added to this mix are the data analytics that popular search engines (e.g., Google and others) can offer regarding data on user website traffic, how many clicks per visit, time spent on each site, and so forth.

Each of these platforms can segment your target market by geography, age, likes and dislikes, etc. There are many platforms to get your company message across to any segment of potential customers. Consequently, companies should develop a strategy to communicate the *value* of their new product using these digital platforms.

It should be noted that people may perceive the same message differently. A strategy should be developed that offers a clear and consistent message. If your product will be advertised on TV, a social media campaign should pick up on the same theme or message. Each channel should reinforce the central message. Why would you want one ad on TV to be different from those ads placed on Facebook or Instagram? Consistency in the message is key.

Placement in Movies and TV

Have you ever noticed the hero of the show sitting down to drink a beer, which is clearly and cleverly placed so that the label can be seen? Today this is a common occurrence where companies pay to have their products prominently displayed on a program or movie. This is a subtle but effective way to use traditional media in a new way.

Marketing Strategy

A strategy should consider all the topics we have discussed, especially Chapter 2 (marketing). Additional steps to a marketing plan:

1. Who is our target market? Where are they?
2. What are our sales goals for the new product?
3. Which media for which demographic group should we use?
4. What is the approved advertising budget?
5. Can we present a link back to our webpage for direct sales?
6. What is our target market? Where are they?
7. How is our product or business unique?
8. How will we collect metrics?

For Small businesses, the internet has been an exceptional opportunity to get your brand and products into the marketplace. Small businesses can get just as much exposure as large companies for very modest costs. It should be acknowledged that small companies do not have a budget that can compete with major corporations. Therefore, social media is the answer.

By far, the most visible social platform is Facebook. Your business should have a Facebook page to build a community and establish relationships. It also allows you to put a face to your brand and describe who you and your company are. It also helps to identify your customer base.

Facebook's Marketplace Ads promote your page right on Facebook, and Facebook provides several tools to help you. Facebook's "Guide to Facebook Ads" will shepherd you through the advertising stages, including planning, creating ads, and interpreting site statistics.

Regardless of whether you use Facebook or not, there are many internet sites to help you promote your product. Others include Pinterest, LinkedIn, and Twitter. There are many others, but these are some of the most popular. It is important that you select the site that best matches your target market. For example, if you are promoting consulting services or new tech products, LinkedIn might be the better option as it is for professional networking.

Your firm should also have a company website that clearly communicates what your company does, and the products or services it offers. Websites have, to a large degree, replaced brochures. Your customers no longer need to collect and store brochures (and many time, brochures are simply trashed), but rather they can go to your company's website and learn about your firm. This also saves your company printing costs.

A company website will allow you the opportunity to post blogs, customer testimonials, and entertaining stories or case studies. You can also collect data analytics about your website visits. Why should you collect data?

Data analytics will enable you to collect data about visits to your web site, which allows you to refine your site content to better serve your customers. The data will likely include the email addresses of people who visit your site. Social media platforms provide data if you have a business account. Google Analytics also offers data related to social media; and more than likely, whoever is your host for your company web page will have data analytics available.

Other marketing targets include local business associations, trade shows, chambers of commerce, and conferences. Recall from Chapter 6 the discussion on business incubators. An incubator will also provide guidance on marketing your product and if connected with an investor group, may also provide advertising funding.

Sales Partner

If may be useful to team with a major retailer to sell your products through the store network. This would include companies like Wal-Mart, Target, Big Lots, and others that have both "bricks and mortar stores" as well as an on-line presence. Clearly, you will have to have the capability to provide large quantities of your product at a discount price.

Other sales partner sites include Amazon and eBay. Using these "stores" will depend to a large extent on whether your target audience frequents those sites and what kind of product you are selling. There are many retail stores that sell on these sites, but each will collect a sales fee and may not be the best alternative, especially if you are selling a groundbreaking new technology or service.

Figure 9.2 shows the process flow to develop a marketing campaign for your new product. Note that the campaign will need to be in line with the marketing strategy and objectives; and may be constrained by the approved budget.

Product Development – Post Review

The purpose of the post product review is *not* an audit. Audit is a word that gives people anxiety. The purpose of the review is to compare the products market acceptance and profitability with the original business parameters. The review should culminate

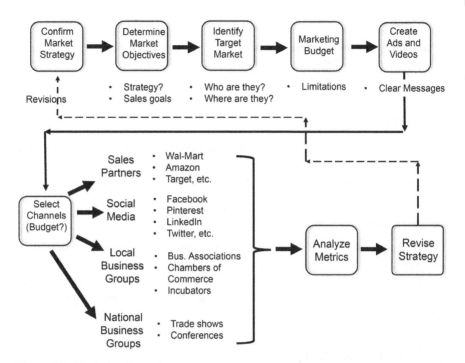

Figure 9.2 Marketing Campaign.

in a formal report that provides an overall evaluation. Further, the report should assist the company in refining how it managed product development – it is an opportunity to improve.

It is the responsibility of the Product Leader to organize, write, and deliver the report. The recipients should include the sponsoring executive and high-level stakeholders. It is appropriate that the Product Leader meet with the product team, suppliers, or other appropriate participants to obtain their input.

Topics to cover in the report may include:

- Introduction – justification
- Overview of the project/product
- Original objectives
- Engineering and Production
- Scope/budget/schedule discussion
- What went right
- What could have been done better
- Identification of outstanding commitments
- Discussion of scope changes
- Marketing strategy and metrics
- Expected competitors
- Next steps: recommendations
- Summary and conclusions.

Introduction

This would include the fundamental reason for developing the new product. Where did the idea originate? What was driving its development? Has market penetration been achieved? Did the project meet its sales targets?

Overview of the Project/Product

Is it profitable or if not, when will it break even (that is, all its development costs are recovered, and going forward, sales are profitable)? How well was it received by customers? Who were the primary participants? (Suppliers, team members, etc.)

Original Objectives

Does the product align with the company's strategic plan? What were the original product objectives and were they realistic? How much did the final configuration (product, idea, or service) differ from the original assumptions? Did the gate reviews help the team stay focused? Were key stakeholders informed of gate review results?

Engineering and Production

What were the challenges in engineering the product? Was the technology feasible? How many design iterations occurred before the final drawings were issued? Was it necessary to purchase new design software (Finite Element Analysis, modeling, etc.)?

Did manufacturing have any issues with prototype or final design? If so, what? Have procedures or manufacturing corrections been required? How effective are these? What changes should occur for the next production run? Is the cost per unit in line with early estimates or those presented in the business case?

Scope/Budget/Schedule Discussion

How many scope changes were required during this product's development? Did any of them decrease costs? Where did most of the change requests originate (engineering, customers, production)?

How close to the original, approved budget was the final costs? Will the product meet the required rate of return?

Did the product get released on time? Where were the primary bottlenecks in meeting schedule? Did our suppliers meet their contractual responsibilities regarding schedule? Should we use them again for Version 2.0?

What Went Right

What methods or techniques were well done that we would want to duplicate with other projects? Were any new processes developed that are applicable elsewhere in the company?

Note: This is an opportunity to take key successes and figure how to apply them to other products or projects. This could be budgeting, communications, working with suppliers, etc. to provide continuous improvement to the corporation.

What Could Have Been Done Better

What areas need improvement? How can future products be developed more quickly? *Note: This is where problems are identified, and solutions offered. This can be engineering problems, excessive production material wastes, or managerial issues.*

Identification of Outstanding Commitments

Are there currently any open contractual commitments? Have all supplier invoices been received and paid? Have engineering final drawings been provided?

Discussion of Scope Changes

How many scope changes were implemented during the product's development? Did any of the scope changes reduce costs? Can any improvements be made to reduce the number of scope changes in the future? What was the total added costs due to scope changes?

Marketing Strategy and Metrics

How effective was the marketing strategy? Did the channels used reach the targeted customers? Have any changes to marketing or advertising occurred due to data metrics (or data analytics)?

Was the approved marketing budget adequate?

Expected Competitors

Have any new competitors emerged with a similar product? How are existing competitors responding to the product? Does this product have a moat? *Note: Moat refers to a castle stronghold that is hard to overcome. Similarly, products can have a moat in the sense that it will be difficult for competitors to challenge or overcome its strong market position.*

Next Steps: Recommendations

Should the company approve and fund Version 2.0? Does the product roadmap layout a reasonable set of enhancements for the next model? What options or alternatives should be considered?

Conclusions

This section simply summarizes the key findings.

Chapter Key Points

- History is littered with products that failed.
- There are several strategies regarding pricing a new product, simply adding on a percentage for profit does not guarantee success or profitability.
- Effective marketing must include social media platforms.
- Data analytics or metrics are important for determining the effectiveness of a marketing campaign. It can provide insights to changing a strategy or customer focus.
- A post launch product review is beneficial for continuous improvement. This allows the company to improve how it designs and produces new products.

Discussion Questions

1. What are the primary advantages to customer input or feedback?
2. Technology problems can be a cause of product failure. However, many times, failure is caused by managerial issues. List a few managerial issues for which team leaders should be aware in order to prevent failure.
3. This chapter discussed different pricing strategies: Customer-focused, profit-focused, competitor-focused, and sales-focused. For the following products, which of these strategies should be used:
   - Laundry detergent
   - A unique software product that will enhance engineering design
   - Custom homebuilder
   - A new electric vehicle battery that is recyclable and will last 500,000 miles.
4. Social media platforms offer companies a broad approach to reaching their target market. What are some of the disadvantages and advantages of these platforms?
5. A sales partner and social media can assist in bringing new products to market. For the previous lists (Question 3), which platforms should be used for marketing:
   - Laundry detergent
   - A unique software product that will enhance engineering design
   - Custom homebuilder
   - A new electric vehicle battery that is recyclable and will last 500,000 miles.
6. The post-review process can potentially take a lot of time and effort to produce. Are there any disadvantages to this "final" report?

Discussion Questions – Answers

1. What are the primary advantages to customer input or feedback?
   Customer input is needed to determine if a new product has any appeal or interest. How can you design a product without customer input?
   Customer feedback is necessary in prototype testing. Did they like the product? Was it easy to use? Is there potential for confusion or product misuse? Customer feedback will help companies determine what changes need to occur in either design or manufacturing.
2. Technology problems can be a cause of product failure. However, many times, failure is caused by managerial issues. List a few managerial issues for which team leaders should be aware to prevent failure.

Managerial failures can occur due to: poor leadership, constant scope changes, poor planning, unrealistic expectations, poor communications, lack of management support, etc.

3. This chapter discussed different pricing strategies: Customer-focused, profit-focused, competitor-focused, and sales-focused. For the following products, which of these strategies should be used:

   - Laundry detergent
   - A unique software product that will enhance engineering design
   - Custom homebuilder
   - A new electric vehicle battery that is recyclable and will last 500,000 miles.

   Laundry Detergent – you may have a new and different laundry detergent, but this product has been around for a long time, has many competitors, and is a commodity. Therefore, you would want a "sales strategy" in which case small margins are acceptable with large volume sales.

   Unique software – this will take time to get market acceptance and strong relationships with customers will help. This could be a combination of customer-focused and sales-focused. The initial release software is always the most expensive, but more sales will quickly reach the break-even point. Having good relationships with customers and an exciting product that will save them time and should also lead to sales.

   Custom homebuilder – while homebuilding is nothing new and there are many competitors, custom homebuilding implies a very customer-focused approach. These homes are not low cost and usually involve significant customer input.

   New electric car battery – The specifications imply a new disruptive technology which any car maker would be pleased to use. This is an instance where a profit strategy may be the most effective (Add a profit percentage to the cost of the product to arrive at a price.).

4. Social media platforms offer companies a broad approach to reaching their target market. What are some of the disadvantages and advantages of these platforms?

   Disadvantages: One must constantly update the web page to keep current and to entice customers to come back. If a sole proprietor, this will take significant time. It may be worthwhile to hire someone to manage the social media accounts. There will be some costs to maintain a business presence on social media.

   Advantages: Offers multiple channels to reach many customers. Platforms can be customized to reach market segments. Metrics are usually provided by the platform for business accounts. Cost is reasonable.

5. A sales partner and social media can assist in bringing new products to market. For the previous lists (Question 3), which platforms should be used for marketing:

   - *Laundry detergent – If direct sales, an online presence like Amazon would be appropriate. If brick and mortar, any of the major retailers (big-box stores and supermarkets) would be useful.*
   - *A unique software product that will enhance engineering design. This will not have broad, general market appeal, but is very specialized. LinkedIn and other professional sites would be appropriate.*
   - *Custom homebuilder – Social media platforms such as Facebook and Pinterest will reach a large target market.*

- *A new electric vehicle battery that is recyclable and will last 500,000 miles. This is a very specialized product but must get the attention of the public as well as car companies. Both general platforms such as Facebook and professional platforms will be appropriate.*

6. The post review process can potentially take a lot of time and effort to produce. Are there any disadvantages to this "final" report?

Without a post review report, how will a company know what the next steps should be? The post review report is meant to help companies continuously improve their business and products.

Notes

1 Campbell, J., The Grocery Store Guy, accessed on September 1, 2021, https://thegrocery storeguy.com/what-is-the-profit-margin-for-grocery-stores.

2 Yahoo.finance.com, accessed on September 2, 2021, https://finance.yahoo.com/quote/KR/key-statistics?p=KR.

Bibliography

Belz, A. (2011). *Product Development*. McGraw-Hill.

Berkun, S. (2010). *The Myths of Innovation*. Sebastopol, CA: O'Reilly Media.

Cooper, R. (2011). *Winning at New Products*. Basic Books.

Norman, D. (2013). *The Design of Everyday Things*. New York, NY.

10

Summary – Connecting the Dots

As stated several times in this book, product development is the future revenues and lifeblood of any company. This justifies the dollars expended through R&D, engineering, and testing of new products, services, and ideas. Ideas start with people and there is no shortage of creative people and companies.

To some extent, a company must nurture an environment where risk-taking and idea-generation are encouraged and rewarded. Some of the most successful companies have been risk-takers and rewarded handsomely for it: Apple, Google, Amazon, and many others. Notably, many traditional firms have also been recognized for their forward thinking and innovation:[1] Ikea, Chick-fil-A, Ford, Toyota, Black & Decker, and John Deere to name a few.

These companies did not get this acknowledgement by luck – it takes a team approach, resources, focus, and a recognition of customer wants or needs. Note that developing new products is risky and requires funding. It would be interesting to know that for every successful product developed by the previously named companies, how many prior attempts have failed? We'll never know the answer of course, but it is probably significant. One highly successful (i.e., profitable) product can make up for several failed ones.

In developing new products, companies really have three choices:

- Develop a completely new, ground-breaking product that no one has tried or thought of before.
- Take an existing product and improve or update it.
- Mimic one of your competitor's products by adding features and making "your" version better, cheaper, more attractive, etc. This is poaching on your competitors' ideas and customers, but this is a common practice.

All products, over time, will find competitors emerging with their own version. This is fundamental to the product lifecycle.

The author is aware of one company that had an interesting strategy. They engineered and manufactured highly technical products, but they knew they would hold a competitive market advantage for at most five years. At that point, their competitors would have reverse-engineered or developed their own competing version. However, during this time, the firm is exceptionally profitable and has a monopoly status on the latest technology in this industry. Once the competitors catch up, the firm would either sell off the product (licensing) or have a new, highly technical product ready to

Product Development: An Engineer's Guide to Business Considerations, Real-World Product Testing, and Launch, First Edition. David V. Tennant.
© 2022 John Wiley & Sons, Inc. Published 2022 by John Wiley & Sons, Inc.

take its place – and provide another five years of market dominance. This was a recognition by executive management that significant investment in R&D was a necessary business expense which paid off handsomely. Did this company experience failure? Yes, many times. But the company's yearly budget dedicated around 12 percent of their net income for R&D and generated multiple patents each year. This is a significant commitment as many firms generally invest 5–7 percent of earnings in R&D.

A Logical Process Flow

As a summary, this section will provide a logical step-by-step approach to developing new products and services. It represents the culmination of this book and should connect-the-dots on the philosophies and techniques presented. It should be noted that using this approach can be very successful but does not guarantee market acceptance for a new product or service. No one can predict market reaction or guarantee market acceptance. Also, some companies, depending on their organization and experience, may find it useful to modify this approach to fit how their company operates. Figure 10.1 Product Development Flow will illustrate visually the order of product development activities described in detail below.

Step 1 – Idea Generation (refer to Figure 2.1).

Before any research, development, testing, or production can occur, it is necessary to generate ideas. Questions to explore include:

- Do our competitors have something similar? Can we improve on it?
- What do our customers want?
- What would be the market potential?
- What is the target market?
- How and where would the product be manufactured?
- Should we initiate a business case or shelve the idea for the time being?

Assuming the idea is accepted, a business case or feasibility study would be initiated.

Step 2 – Taking the Business Case from Concept to Reality (refer to Chapter 5 and Figure 5.1).

The business case will take a hard look at the feasibility of pursuing the idea. It will attempt to determine investment costs, a timeline, suggest alternatives, profitability (i.e., NPV – Net Present Value), potential sales, and make a recommendation. Should the idea pass the necessary criteria, senior management may approve the "project." A preliminary budget and schedule may be provided at this time. Assuming the potential product is approved, the following steps would occur. Additional relevant figures are 2.6 and 9.2; chapters two and nine, respectively.

Step 3 – Customer Research (refer to Figure 5.4)

- Marketing performs customer focus groups and seeks input from suppliers and others (some companies may perform focus groups during the development of the business case). The Product Manager is assigned to project.
- Begin planning for sales and marketing strategy: product, promotion, placement, etc.
- Begin developing preliminary cost and pricing models (refer to business case).

- Stakeholders are identified.
- First gate review.

Step 4 – Initiation

Engineering and Marketing are performing simultaneous work on parallel paths and must communicate constantly. Marketing will provide customer feedback and focus group results so the product can meet or exceed customer expectations. Engineering will take this information and incorporate into the product's design. The steps include:

- Assemble the project team and confirm team leader (product manager).
- Study the business case and customer feedback (from marketing group).
- Develop the vision statement (Figure 5.6).
- Assign roles and responsibilities for each team participant.
- Confirm project objectives with key stakeholders.
- Develop list of project assumptions and constraints.
- Determine preliminary list of required resources.
- Perform initial risk review. See sample risk review, Table 8.4 Qualitative Risk Review.
- Determine where product will be manufactured: domestic or offshore, in-house or contract?
- Second gate review.

Step 5 – Pre-Design This phase involves working with Finance and Manufacturing *and is a full team activity*.

Depending on the product/project complexity, this may take significant time. Items generated during this time include:

- A detailed cost estimate, cash flow, WBS (Work Breakdown Structure) and schedule. See sample WBS, Figure 5.2.
- Product manager works with accounting and finance to determine, based on cost estimate, a proforma, NPV, pricing, and breakeven point (Figure 2.5). This is an ongoing activity with periodic updates.
- Engineering works with manufacturing to determine process flows, assembly techniques, types of production setup, etc.
- Quality group is brought on board to set up QA/QC program.
- Develop and issue formal project plan (See Chapter 5 Project Plan Table of Contents).
- Supplier delivery timelines – develop procurement plan and desired delivery schedule.
- Perform risk review.
- Manage scope and proposed changes. Refer to Figure 8.6 Change Request Review Process.
- Third gate review.

Step 6 – Detailed Planning: Design

- Begin engineering design (with R&D, if appropriate) – Figure 3.2.
- Engineering conducts multiple design review meetings (which engineering disciplines?).

- Specifications are developed and potential suppliers are identified.
- Supply chain becomes involved in the procurement process. Refer to Figure 4.2 Procurement Process.
- Supply chain works with engineering to develop an RFP (request for proposal).
- RFP is released to potential suppliers.
- Proposals from suppliers are evaluated and contracts are awarded.
- Perform risk review.
- Fourth gate review.
- Apply for patent? (Fig 6.3)

Based on supplier quotes and the determined scope, the cost estimate or budget will need to be refined. As the project gets closer to production, the costs will evolve and become clearer.

It is entirely possible that R&D will be occurring at the same time; or, the original idea may have been devised in the R&D department, at which point there may already be some progress. If there is engineering design involved, it happens during the planning phase along with developing and testing prototypes.

Step 7 – Testing and Review: Marketing and Engineering groups communicate with customers to refine final design.

- Engineering develops prototype based on customer desires and use of software modeling.
- Continued financial review of costs, NPV, breakeven point, etc.
- Confirmation of end user requirements.
- Prototype goes through testing. This may involve customer tests and reviews.
- Design group evaluates new product from a human engineering perspective. See Chapter 3, Ergonomics (human factors).
- Prototype evaluation for accidental or intentional misuse. Refer to Figure 3.8 Product Misuse Review (legal and engineering review).
- Manage scope.
- Design review.
- Simultaneous to engineering design, the marketing group is developing promotional materials and final pricing guidelines. Refer to Figure 9.2 Marketing Campaign.
- Perform risk review.
- Fifth gate review.

Step 8 – Execution

- Engineering, Manufacturing/Production, Marketing, and Quality groups coordinate to issue new product.
- Final design review.
- Prototype data and final input from marketing is used to finalize drawings.
- Risk review.
- Gate review.
- Final design drawings are issued to manufacturing facility "Approved for Production."
- Supplier deliveries arrive (raw materials, components, etc.).

- Manufacturing produces first small batch to test for quality and that product is meeting performance and quality standards.
- Full production commences.
- Marketing issues promotional activities, commercials, social media, etc.
- Product is shipped to distributors/customers.

Step 9 – Closing

- Final invoices from suppliers or contractors are paid.
- Excess materials are stored or disposed.
- Post product review (customer feedback included).
- Celebrate success.
- Team members reassigned.
- Product roadmap consulted for planning next product iteration. Refer to Figure 6.2 Product Roadmap.
- Close contracts and project accounts.

The above discussion is a structured approach to product development, design, and production. A company may decide to modify this approach to suit their requirements or company culture. A small company or entrepreneur may have a shortened version of this process; however, most of these steps should proceed in this order with minimal shortcuts to ensure success.

Figure 10.1 Product Flow Summary illustrates visually the nine steps discussed above.

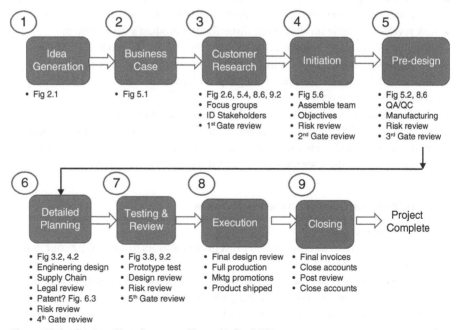

Figure 10.1 Product Flow Summary. Figure by David Tennant

The previous discussion and Figure 10.1 should illustrate the importance of leadership to the product team and strong communications between the functional areas. There are a number of functional areas involved in our above discussions. These are represented:

- R&D/Engineering
- Supply Chain
- Marketing
- Manufacturing
- Executive management (stakeholders)
- Suppliers
- Legal
- Warehouse (receiving supplier deliveries)
- Finance/Accounting.

Remember that projects generally fail due to managerial issues. While technical problems certainly appear, these are less common. Assigning the right person to lead a product development effort is critical. No less important are the team members assigned to the project and selected suppliers. It takes a team effort of people from a variety of functional areas to succeed. Remember, the product leader cannot do it all. Please note that the process described in this chapter is robust; however, some companies may decide that a variation of this, or a different sequence of activities, will be more appropriate. The product development process must complement the firm's organizational structure and internal policies.

Chapter Key Points

- Product development, like project management, should use a structured approach.
- Companies have three choices in product development: new, improve, or copy.
- Risk reviews and gate reviews occur throughout a product's development.
- Customer or end user input should be accepted throughout the product's development.
- The costs estimate will change and evolve throughout the product's timeline.
- There are three distinct reviews during a product's development: risk reviews, gate reviews, and design reviews.
- Effective communications and leadership are essential to success.
- Before starting full production, a small subset of product should be produced, inspected and tested to ensure quality and functional goals are being met. Once this has occurred, and the tests are successful, full production can begin.

Summary Discussion Questions (Covering all chapters)

1. Why is a product roadmap useful?
2. What techniques can be employed to prevent competitors from stealing your design or production secrets?
3. If a proposed budget or cost estimate is contained in the business case, why do we want to develop another estimate during the planning phase?
4. Rather than have Supply Chain issue RFPs and evaluate proposals, why doesn't engineering perform this function since most contracts will be dealing with technical issues?
5. What is the purpose of the project/product post review?

6. The discussions in this chapter highlight the depth and time-consuming aspect of product development. Small companies may not have all of the expertise to follow this structure. Is there anything that small firms can do to shorten this process?
7. Since leadership is an important quality, how can I improve my leadership abilities?
8. Why is a cash flow diagram important?
9. Why is developing a vision statement so important? Can't we just formulate objectives and work from that?
10. Why are multiple risk reviews performed during a product's development?

Summary Discussion Questions – Answers

1. Why is a product roadmap useful?
   The product roadmap describes what the future features and functionality of the product's next revision will include. Many times, it is not feasible to build all possible features into the initial product. The reasons may include new technology making this product obsolete, putting all the requested features into the initial product may take too long and competitors would beat you to market. The next release will incorporate improvements based on customer experience with Version 1.

2. What techniques can be employed to prevent competitors from stealing your design or production secrets?
   - *One can always seek a patent (but this offers no firm guarantee unless you are willing to legally pursue those who steal your idea or product).*
   - *You can do all of the development, design, and production in-house.*
   - *Ensure that your suppliers and contract manufacturers are ethical and willing to sign a non-disclosure agreement.*

3. If a proposed budget or cost estimate is contained in the business case, why do we want to develop another estimate during the planning phase?
   A business case, while many times is very detailed, just does not have enough information to make a definitive estimate. A cost estimate is developed during the detailed planning stage, which happens after the business case:
   - *Supplier quotes usually come later in the timeline. There is no way a business case can gather supplier quotes without a well-defined scope, which also happens in the planning stage.*
   - *It will take the engineering team time to assign all known costs in a product estimate. How would the business case team know how much time the design effort will take (and therefore the costs)?*

4. Rather than have Supply Chain issue RFPs and evaluate proposals, why doesn't engineering perform this function since most contracts will be dealing with technical issues?
   - *A publicly held company (i.e., listed on the stock exchange), legally must keep purchasing approvals separate from other departments. It would be illegal for the same person (say, an engineer) to approve a purchase order, a receiving report (warehouse), and approve the invoice.*
   - *While the engineering team would clearly understand the technical requirements more clearly, Supply Chain personnel are very good at negotiating prices, developing*

contracts, and performing contract administration. This takes a large burden off the engineering group and allows them to focus on what they do best: design.

5. What is the purpose of the product/project post-review?
   - *To determine what activities went well during a product's development. It may be that some of the practices can assist other project teams be successful by adopting them.*

6. The discussions in this chapter highlight the depth and time-consuming aspect of product development. Small companies may not have all the expertise to follow this structure. Is there anything that small firms can do to shorten this process?
   - *Some activities can be outsourced to consultants or contract manufacturers.*
   - *There is a long list of structured activities in product development. It is possible, depending on the complexity of a product, that some of these activities can be downsized.*

7. Since leadership is an important quality, how can I improve my leadership abilities?
   - *Read books and attend seminars on leadership.*
   - *Observe how others in high level positions carry themselves.*
   - *Never stop learning.*
   - *Talk with leaders and see if they will mentor you.*
   - *Honor commitments.*
   - *Ensure that you exhibit a high level of integrity and trust.*

8. Why is a cash flow diagram important?
   - *From a budgeting standpoint, companies like to know how their expenditures will be spread over the fiscal year.*
   - *Some projects may cross over into the next year's budget cycle. It is important to know how much funding your project needs this year and next.*
   - *If you plan to perform earned value tracking on your project, cash flows are necessary (Note: earned value was not discussed in this book but is a method to assess project progress for both budget and schedule.).*

9. Why is developing a vision statement so important? Can't we just formulate objectives and work from that?
   - *Typically, the vision statement is the guiding overall statement.*
   - *The objectives should support the vision statement not the other way around.*
   - *The vision statement allows senior level people to understand at a glance what the project or product is about. Objectives, which tend to be more specific and at a lower level do not always clearly delineate the overall mission.*

10. Why are multiple risk reviews performed during a product's development?

    As the project proceeds, risks identified early in the project will go away, but new ones will emerge as one goes through the product's development.

    For example, we may decide that contractor performance is a risk. However, once a reliable contractor has been placed under contract, that risk will dissipate, but new ones may emerge regarding weather, material shortages, or labor issues.

Reference

1 https://americaninnovationindex.com/aii-conference.

Index

Note: Page numbers followed by "*f*" denotes figure and "*t*" denotes table respectively.

Product Development: An Engineer's Guide to Business Considerations, Real-World Product Testing, and Launch, First Edition. David V. Tennant.
© 2022 John Wiley & Sons, Inc. Published 2022 by John Wiley & Sons, Inc.